T0289673

# Theory of Machines: Kinematics and Dynamics

# Theory of Machines: Kinematics and Dynamics

## Dustin Green

New York

Published by NY Research Press
118-35 Queens Blvd., Suite 400,
Forest Hills, NY 11375, USA
www.nyresearchpress.com

Theory of Machines: Kinematics and Dynamics
Dustin Green

International Standard Book Number: 978-1-64725-427-8 (Hardback)

**Cataloging-in-Publication Data**

Theory of machines : kinematics and dynamics / Dustin Green.
    p. cm.
Includes bibliographical references and index.
ISBN 978-1-64725-427-8
1. Machinery, Kinematics of. 2. Mechanics, Analytic. 3. Kinematics. 4. Dynamics.
5. Mechanical engineering. I. Green, Dustin.
TJ175 .T44 2023
621.811--dc23

# Contents

Preface ................................................................................................ VII

Chapter 1  **Mechanisms and Kinematic Analysis** .................................................... 1

    1.1 Mechanisms: Basic Kinematic Concepts     1

    1.2 Degrees of Freedom for Plane Mechanism     8

    1.3 Four Bar Chains and their Inversions     14

    1.4 Kinematic Analysis     24

    1.5 Velocity in Four Bar Mechanism     40

Chapter 2  **Inertia Forces in Reciprocating Parts** ................................................ 51

    2.1 Reciprocating Engine     51

    2.2 Dynamically Equivalent System     75

    2.3 Turning Moment Diagrams for different Types of Engines     85

    2.4 Friction of a Screw: Nut and Square Threaded Screw     95

Chapter 3  **Brakes, Dynamometers and Gear Trains** ........................................ 118

    3.1 Brakes and its Classification     118

    3.2 Dynamometer: Absorption and Transmission     132

    3.3 Gear Trains     138

    3.4 Belt, Rope and Chain Drives     150

**Permissions**

**Index**

# Preface

This book has been an outcome of determined endeavour from a group of educationists in the field. The primary objective was to involve a broad spectrum of professionals from diverse cultural background involved in the field for developing new researches. The book not only targets students but also scholars pursuing higher research for further enhancement of the theoretical and practical applications of the subject.

A machine refers to a physical system, which utilizes power for applying forces and regulating movement in order to complete an action. The theory of machines is an engineering science discipline that studies the relative motion of several machine parts as well as the forces acting on them. It is categorized into four branches, namely, kinetic, kinematics, statics, and dynamics. Kinematics is a branch of physics, which defines the motion of bodies along with systems of bodies without taking into account the forces responsible for moving them. It is often referred to as the geometry of motion. Dynamics as a branch of classical mechanics studies forces and their impact on motion. This book is compiled in such a manner, that it will provide in-depth knowledge about the theory of machines. It will help the readers in keeping pace with the rapid changes in this area of study.

It was an honour to edit such a profound book and also a challenging task to compile and examine all the relevant data for accuracy and originality. I wish to acknowledge the efforts of the contributors for submitting such brilliant and diverse chapters in the field and for endlessly working for the completion of the book. Last, but not the least; I thank my family for being a constant source of support in all my research endeavours.

**Dustin Green**

# Mechanisms and Kinematic Analysis

## 1.1    Mechanisms: Basic Kinematic Concepts

### Mechanism

If a motion of any of movable links results in definite motion of the others, the linkage is known as mechanism.

### Machine

A machine may be defined as a device consisting of fixed and moving parts that modifies mechanical energy and transmits it into a more useful form.

Mechanisms are classified into six basic types:

- Screw Mechanisms.
- Wheel mechanisms (roller mechanisms).
- Cam mechanisms.
- Crank mechanisms.
- Belt mechanisms.
- Ratchet and lock mechanisms.

### Difference Between Mechanism and Structure

| S.No. | Mechanism | Structure |
|---|---|---|
| 1. | Mechanism transmits and modifies motion. | No relative motion exists between its members. |
| 2. | A mechanism is the skeleton outline of the machine to produce definite motion between various links. | It does not convert the available energy into work. |
| 3. | Example: Clock work, Typewriter. | Example: Shaper and Lathe. |

## Kinematic Pair Depending upon Nature of Contact

### 1. Lower Pair

If a pair in motion has a surface contact between the two elements. It is called a lower pair.

Example: Nut and bolt.

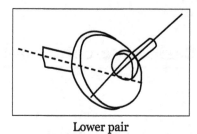

Lower pair

### 2. Higher Pair

If a pair in motion has a line or point contact between the two elements, it is called a higher pair.

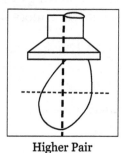

Higher Pair

## Kinematic Pairs Depending upon Relative Motion

### 1. Sliding Pair

When the elements have a motion relative to each other it is known as sliding pair.

Sliding Pair

### 2. Turning Pair

When two elements are connected such that one element revolves around the other. It forms a turning pair.

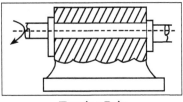
Turning Pair

## 3. Rolling Pair

When one element is free to roll over the other it forms a rolling pair.

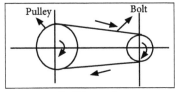
Rolling pair

## 4. Screw Pair (Helical Pair)

In, screw pair, one link is connected to have a combination of turning and sliding motion relative to another element.

Screw Pair

## 5. Spherical Pair

In a spherical pair, one link is constrained to swivel in or about the other fixed point.

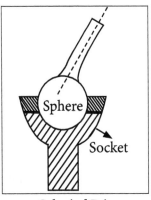
Spherical Pair

## Kinematics Pairs Depending upon Mechanical Arrangements for Constraining Motion

### 1. Closed Pair (Self-Closed Pair)

When two elements of a pair are held together mechanically, they constitute a closed pair. All the lower pairs are self-closed pair.

### 2. Unclosed Pair (Open Pair (or) Force-Closed Pair)

When two links or elements are not held together mechanically, they constitute an unclosed pair.

Example: Cam and Follower.

### Links and its Classification

### Link

A single part (or an assembly of rigidly connected parts) of a machine, which is a resistant body having a relative motion with the other parts of machine is known as a Link or Element.

Based on the rigidity, links are classified into:

### 1. Rigid Link

Rigid links are those links that does not undergo any change of shape when they transmit motion. In reality, no rigid links exist. But kinematic links whose deformation is very small are considered as rigid links. These links do not undergo significant deformation while transmitting motion.

Examples: Crankshafts, connecting rods and cam followers.

### 2. Flexible Link

A flexible link undergoes partial deformation while transmitting motion. Its deformation does not affect its transmission effectiveness.

Examples: Flexible links are belts (in belt drives) and chains (in chain drives).

### 3. Fluid Link

A fluid link makes use of a fluid (liquid or gas) to transmit motion, by means of pressure. Fluid links undergo deformation while transmitting motion.

Examples: Pneumatic punching presses, hydraulic jacks and hydraulic brakes.

## 1.1.1 Classification of Kinematic Pairs

## Kinematic Pair

When any two links or Elements are connected in such a way that their relative motion is completely or successfully constrained, they form a Kinematic Pair.

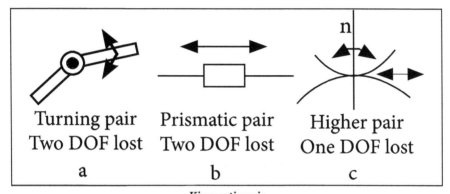

Kinematic pair

Classification of pairs is based on the following considerations:

1. Nature of contact between them (i.e., type of contact).

2. Nature of Relative motion between them (i.e., type of relative motion).

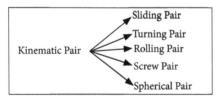

3. Nature of Mechanical arrangement for constraining relative motion (i.e, type of constraint).

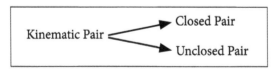

## Lower Pair and Higher Pair

When the two elements of a pair have a surface contact when relative motion takes place and surface of one element slides over the surface of the other, the pair formed is known as lower pair.

When the two elements of a pair have a line (or) point contact when relative motion take places and the motion between the two elements is partly turning had partly sliding, then the pair is known as higher pair.

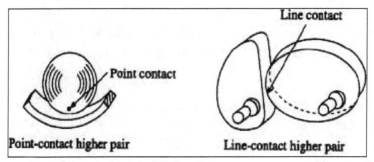

Higher Pair

## Translation and Rotation

1. Translation: A body has translation if it moves so that all straight lines in the body move to parallel positions. Example link 4 in the above figure.

2. Rotations: In rotation, all points in a body remains at fixed distances from a line which is perpendicular to the plane of rotation. This line is the axis of rotation and points in the body describing circular paths about it. E.g., link 2 in the above figure.

### 1.1.2 Degrees of Freedom

The number of input parameters which must be independently controlled in order to bring the mechanism into a useful engineering purpose.

The degrees of freedom of a pair is defined as the number of independent relative motions, each translational and rotational, a pair can have.

Degrees of freedom = 6 − no. of restraints.

To find the number of degrees of freedom for a plane mechanism we have an equation known as Grubler's equation and is given by,

$$n = 3(l - 1) - 2j - h$$

$$l = 3, j = 2, h = 1$$

$$n = 3(3 - 1) - 2(2) - 1$$

$$n = 1$$

If,

- n > 0, Results in a mechanism with 'n' degrees of freedom.
- n < 0, Results in a statically indeterminate structure.
- n = 0, Results in a statically determinate structure.

## 1.1.3 Kinematic Chain, Binary, Ternary and Quaternary Joints and Links

A kinematic chain is arrangement of kinematic pairs in such a way that each link forms a part of two pairs and the motion of each is relative to other is definite and the last link is joined to the first link to transmit definite motion.

Example: Beam Engine and slider crank mechanism.

Joint can be classified into:

- Binary joint.

- Ternary joint.

- Quarternary joint.

### Binary Joint

A joint is known as binary joint, if two links are joined at the same connection. Figure (a) shows a chain having four links 1, 2, 3, 4 and four joints A, B, C and D. At each joint, two links are connected. Hence these joints are binary joints.

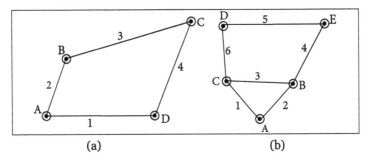

(a)                                             (b)

### Ternary Joint

A joint is known as ternary joint if three links are joined at the same connection. Figure (b) shows a chain having six links and five joints. At joints C and B, three links are connected and hence these joints are ternary joints. But joints A, D and E are binary joints.

### Quaternary Joints

A joint is known as quaternary joints if four links are connected at the same connection.

Link can be classified into:

- Binary link

- Ternary link

- Quarter-nary link

A link, to which two links are connected, is known a binary link. In figure (a), the links 1, 2, 3 and 4 are binary links, as to each link two links are connected. For example to link 1, link 2 and link 4 are connected. Similarly to link 2, link 3 and link 1 is connected.

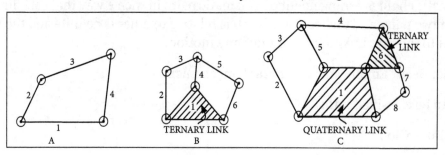

A link, to which three links are connected, is known a ternary link. In figure (b), the link 1 is a ternary link, as to link 1, three links namely 2, 4 and 6 are connected. Link 6 in figure (c) is also a ternary link, as three links (i.e. link 4, link 7 and link 1) are connected to link 6.

A link, to which four links are connected, is known a quaternary link. In figure (c), the link 1 is a quaternary link, as to link 1, four links namely link 2, link 5, link 6 and link 8 are connected.

## 1.2  Degrees of Freedom for Plane Mechanism

The number of degrees of freedom of a mechanism is also called the mobility of the device. The mobility is the number of input parameters (usually pair variables) that must be independently controlled to bring the device into a particular position. The Kutzbach criterion, which is similar to Gruebler's equation, calculates the mobility.

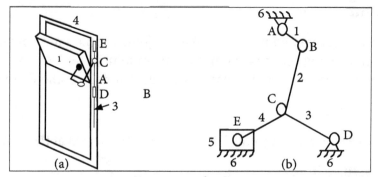

Transom mechanism

In order to control a mechanism, the number of independent input motions must equal the number of degrees of freedom of the mechanism. An example, the transom in above the figure(a) has a single degree of freedom, so it needs one independent input motion to open or close the window. That is, we just push or pull rod 3 to operate the window.

To see another example, the mechanism in above the figure(b) also has 1 degree of freedom. If an independent input is applied to link 1 (e.g., a motor is mounted on joint A to drive link 1), the mechanism will have the prescribed motion.

## Working of Kutzbach Criterion

The mobility of a mechanism is defined as the number of input parameters (usually pair variables) which must be controlled independently in order to bring the device into a particular position.

It is possible to express the number of degrees of freedom of a mechanism in terms of the number of links and the number of pair connections of a given type. This is known as number synthesis.

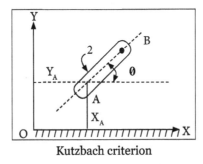
Kutzbach criterion

Let there are two links 1 and 2 in which link 1 is fixed, as shown in figure. The link 2 has a point A over it and translated by co-ordinates xa and Ya.

It can be written as $A(X_a, Y_a)$.

A and B makes an angle $\theta$ with the fixed link 1 (OX). Link 2 specified by two variable $(X_a, Y_a,)$.

Let, Number of Links = l.

So, Number of Movable links = (l - 1) and total number of degree of freedom before they are connected to any other link = 3(l - 1).

If, J = Number of binary Joints or lower pair,

h = Number of higher pairs,

$n = 3(l - 1) - 2j - h$.

This equation is called Kutzbach criterion for the movability of a mechanism having plane mechanism.

### 1.2.1 Gruebler's Criterion

The degrees of freedom of a mechanism is the number of independent relative motions among the rigid bodies. An example, figure given below shows several cases of a rigid body constrained by different kinds of pairs.

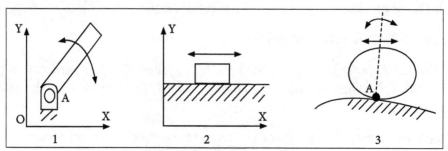

Rigid bodies constrained by different kinds of planar pairs.

In the above figure (1), a rigid body is constrained by a revolute pair which allows only rotational movement around an axis. It has one degree of freedom, turning around point A. The two lost degrees of freedom are translational movements along the X and Y axes. The only way the rigid body can move is to rotate about the fixed point A.

In the above figure (2), a rigid body is constrained by a prismatic pair which allows only translational motion. In two dimensions, it has one degree of freedom, translating along the X axis. In this example, the body has lost the ability to rotate about any axis and it cannot move along the Y axis.

In the above figure (3), a rigid body is constrained by a higher pair. It has two degrees of freedom i.e., translating along the curved surface and turning about the instantaneous contact point.

In general, a rigid body in a plane has three degrees of freedom. Kinematic pairs are constraints on rigid bodies that reduce the degrees of freedom of a mechanism. These pairs reduce the number of the degrees of freedom. If we create a lower pair figure (1),(2), the degrees of freedom are reduced to 2. Similarly, if we create a higher pair figure (3), the degrees of freedom are reduced to 1.

Therefore, we can write the following equation:

$$F = 3(n-1) - 2l - h$$

Where,

l = Number of lower pairs (one degree of freedom).

F = Total degrees of freedom in the mechanism.

n = Number of links (including the frame).

h = Number of higher pairs (two degrees of freedom).

## 1.2.2 Inversion of Mechanism

In this mechanism, the link AC (i.e., link 3) forming the turning pair is fixed, as shown in figure.

The link 3 corresponds to the connecting rod of a reciprocating steam engine. The driving crank CB revolves with uniform angular speed about the fixed center C. A sliding block attached to the crank pin at B slides along the slotted bar AP and thus causes AP to oscillate about the pivoted point A.

A short link PR transmits the motion from AP to the ram which carries the tool and reciprocates along the line of stroke $R_1R_2$. The line of stroke of the ram (i.e., $R_1R_2$) is perpendicular to AC produced.

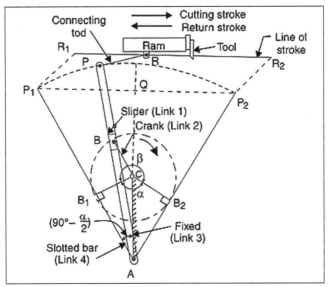

Crank and slotter lever quick return mechanism.

In the extreme positions, $AP_1$ and $AP_2$ are tangential to the circle and the cutting tool is at the end of the stroke. The forward or cutting stroke occurs when the crank rotates from the position $CB_1$ to $CB_2$ (or through an angle) in the clockwise direction. The return stroke occurs when the crank rotates from the position $CB_2$ to $CB_1$ (or through angle $\alpha$) in the clockwise direction. Since the crank has uniform angular speed,

$$\frac{\text{Time of cutting stroke}}{\text{Time of return stroke}} = \frac{\beta}{\alpha} = \frac{\beta}{360° - \beta} \text{ or } \frac{360° - \alpha}{\alpha}$$

Since the tool travels a distance of $R_1R_2$ during cutting and return stroke, therefore travel of the tool or length of stroke is given by,

$$= R_1R_2 = P_1P_2 = 2\,P_1Q = 2\,AP_1 \sin P_1AQ$$

$$= 2AP_1 \sin\left(90° - \frac{\alpha}{2}\right) = 2\,AP\cos\frac{\alpha}{2} \cdots (\because AP_1 = AP)$$

$$= 2AP \times \frac{CB_1}{AC} \cdots \left(\because \cos\frac{\alpha}{2} = \frac{CB_1}{AC}\right)$$

$$= 2\,AP \times \frac{CB}{AC} \cdots (\because CB_1 = CB)$$

Therefore the return stroke is completed within shorter timer. Thus it is called quick return motion mechanism.

## Expression for Length of Stroke

In this mechanism, the link AC (link 3) forming the turning pair is fixed. The driving crank CB revolves with uniform angular speed about the fixed center C. A sliding block attached to the crank pin at B slides along the slotter bar AP and thus causes AP to oscillate about the pivoted point A.

A short link PR transmits the motion from AP to the ram which carries the tool and reciprocates along the line of stroke $R_1 R_2$. In the extreme positions, $AP_1$ and $AP_2$ are tangential to the circle and the cutting tool is at the end of the stroke. The forward stroke occurs when the crank rotates from the position $CB_1$ to $CB_2$ is the clockwise direction. The return stroke occurs when the crank rotates from the position $CB_2$ to $CB_1$ is clockwise direction. Since the crank has uniform angular speed, therefore,

$$\frac{\text{Time of cutting stroke}}{\text{Time of return stroke}} = \frac{\beta}{\alpha}$$

$$= \frac{\beta}{360 - \beta} \ \ (\text{or}) \ \ \frac{360 - \alpha}{\alpha}$$

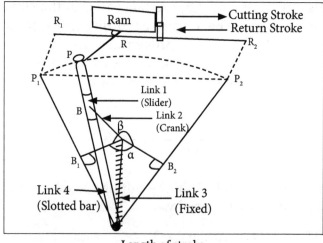

Length of stroke

Since the tool travels a distance of $R_1 R_2$ during cutting and Return stroke, therefore travel of the tool or length of stroke.

$$= R_1 R_2 = P_1 P_2 = 2P_1 Q$$

$$= 2AP_1 \sin\left[90 - \frac{\alpha}{2}\right]$$

$$= 2AP \cos\frac{\alpha}{2}$$

$$= 2AP \cdot \frac{CB_1}{AC}$$

$$= 2AP \times \frac{CB}{AC}[CB_1 = CB]$$

## Problems

The transom above the door is shown in the figure (a). The opening and closing mechanisms are shown in figure (b). Let us calculate its degree of freedom.

Solution:

Given:

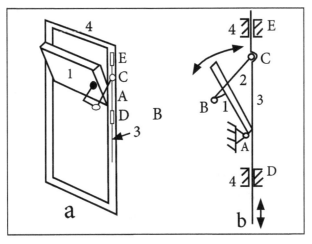

n = 4 (link 1, 3, 3 and frame 4)

l = 4 (at A, B, C, D)

h = 0

F=3(n-1) - 2l - h

F= $3(4-1) - 2 \times 4 - 1 \times 0 = 1$

Note: D and E function as a same prismatic pair, so they only count as one lower pair.

## 1.3    Four Bar Chains and their Inversions

Important inversion of four bar chain are:

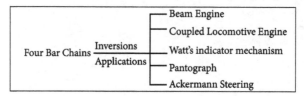

This mechanism is used to convert the rotary motion into reciprocating motion.

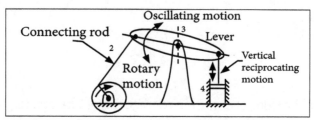

Inversion of four bar chain.

### Double Crank Mechanism (Second Inversion)

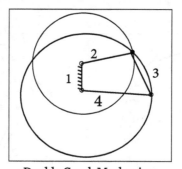

Double Crank Mechanism

If the shortest link, i.e., link 1 (crank) is fixed, the adjacent links 2 and 4 would make complete revolutions, as shown in figure. The mechanism thus obtained is known as crank-crank or double crank mechanism or rotary-rotary converter.

### Coupling Rod of a Locomotive

This is an example of a double crank mechanism where both cranks rotate about the points in the fixed link. It consists of four links. The opposite links are equal in length, since links 1 and 3 work as two cranks, the mechanism is also known as rotary-rotary converter.

### Crank and Lever Mechanism (First Inversion)

As shown in figure (a), link 1 is the crank, link 4 is fixed and link 3 oscillates where-as in figure (b), link 2 is fixed and link 3 oscillates. The mechanism is also known as crank-rocker mechanism or a crank-lever mechanism or a rotary-oscillating converter.

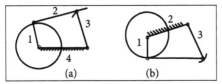

Crank and Lever Mechanism.

## Applications

Beam Engine:

This is an example of crank-lever mechanism, where one link oscillates, while the other rotates about the fixed link, as shown in figure below.

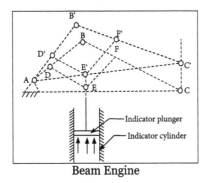

Beam Engine

Watt's engine indicator (double lever mechanism) Pantograph:

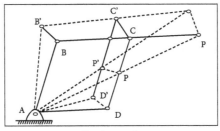

Pantograph

Pantograph is a device which is used to reproduce a displacement exactly in an enlarged or reduced scale. It is used in drawing offices, for duplicating the drawings, maps, plans, etc. As shown in figure, it is a four bar mechanism in the form of a parallelogram ABCD with link BC extended to P.

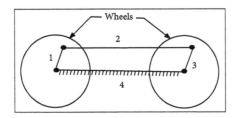

This mechanism is used to transmit rotary motion from one wheel to the other wheels.

## Third Inversion

If the link opposite to shortest link is fixed, i.e., link 3 fixed, then the shortest link (link 1) is made coupler and the other two links 2 and 4 would oscillate as shown in figure. The mechanism thus obtained is known as rocker-rocker or double-rocker or double-lever mechanism or an oscillating-oscillating converter.

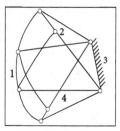

Third Inversion

## Applications

- Watt's indicator mechanism.

- Pantograph.

- Ackermann steering.

### Watt's Indicator Mechanism

This mechanism was invented by watt for his steam engine to guide the piston rod. It is also known as simplex indicator.

It consists of four links: Fixed link at A, link AB, link BC and link DEF, connected to the piston of the indicator cylinder. Links BC and DEF work as levers and due to this, the mechanism is also known as double lever mechanism. The displacement of the lever DEF is directly proportional to the steam or gas pressure in the indicator cylinder.

In figure, full lines depict the initial position of the mechanism, whereas the dotted line shows the position of the mechanism when gas or steam pressure acts on the indicator plunger.

From the similar triangles $A_p$ D and $C_p$ P, it is clear that p and P will follow similar path. Ratio of the motion of p to the motion of P will be,

$$\frac{A_p}{AP} \text{ or } \frac{BC}{BP}$$

### Ackermann Steering

The Ackermann steering mechanism for a motion car is shown in figure. In Ackermann steering, the mechanism ABCD is a four bar chain. The shorter links AB and DC equal in length and the longer links AD and BC are of unequal length.

When the vehicle moves along a straight path, the longer links AD and BC remain parallel as shown in figure (a). When the vehicle is steering to the left (or right), the piston of the mechanism is shown in figure (b). The length of the links are so proportioned that the lines drawn from the axes of all the four wheels intersect at a common point $p_1$. This fact ensures that the relative motion between the types and the road surface is pure rolling.

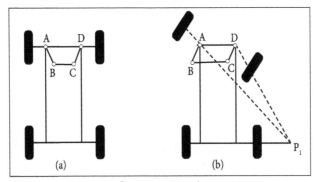

Ackermann Steering

## Sliding Connectors

Sliding connectors are used when one slider (the input) is to drive another slider (the output). Usually the two sliders operate in the same plane but in different directions.

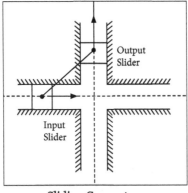

Sliding Connectors

The figure shows a sliding connector which is obtained by a rigid link pivoted at each of a slider. This is called as double slider crank mechanism.

## Inversions of Slider Crank Mechanism

Kinematic Inversion: The process of fixing different links of a kinematic chain one at a time to produce distinct mechanisms is called kinematic inversion. Here the relative motions of the links of the mechanisms remain unchanged.

The Inversion of Slider Crank Chain: A slider-crank chain has the following inversions.

## First Inversion

This inversion is obtained when link 1 is fixed and links 2 and 4 are made the crank and the slider respectively shown in figure (1(a)).

Applications: Reciprocating engine, Reciprocating compressor.

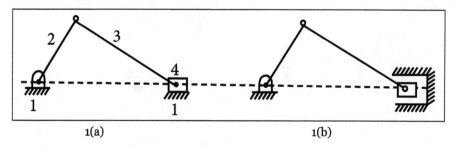

1(a)                                              1(b)

As shown in figure (1(b)), if it is a reciprocating engine, 4 (piston) is the driver and if it is a compressor, 2 (Crank) is the driver.

## Second Inversion

Firing of link 2 of a slider-crank chain results the second inversion.

The slider-crank mechanism of figure (1(a)) can also be drawn as shown in figure (2(a)). Further, when its link 2 is fixed instead of link 1, link 3 along with the slider at its end B becomes a crank. This makes link 1 to rotate about 0° along with the slider which also reciprocates as shown figure (2(b)).

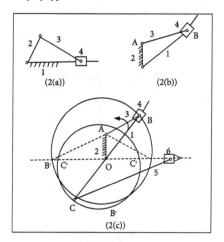

## Whit Worth Quick-Return Mechanism

It is a mechanism used in workshops to cut metals. The forward stroke takes a little longer and cuts the metal whereas the return stroke is idle and takes a shorter period.

Slider 4 rotates in a circle about A and slides on link 1 figure (2(c)). C is a point on link 1 extended backwards where link 5 in pivoted. The other end of link 5 is pivoted to the tool, the forward stroke of which cuts the metal. The axis of motion of slider 6 (tool) passes through 0 and is perpendicular to OA, the fixed link. The crank 3 rotates in the counter clockwise direction.

Initially, let the slider 4 be at B' so that C be at C'. Cutting tool 6 will be in the extreme left position. With the movement of the crank, the slider traverses the path B'BB" whereas point C moves through C'CC". Cutting tool 6 will have the forward stroke. Finally, slider B assumes the position BB" and cutting tool 6 is in the extreme right position. The time taken for the forward stroke of slider 6 is proportional to the obtuse angle B"AB' at A.

Similarly, slider 4 completes the rest of the circle through path B"B"'B' and C passes through C"C"'C'. There is backward stroke of tool 6. The time taken in this proportional to the acute angle B"AB' at A.

Let, $\theta$ - obtuse angle B'AB" at A.

$\beta$ - acute angle B'AB" at A.

Then,

$$\frac{\text{Time of Cutting}}{\text{Time of Re turn}} = \frac{\theta}{\beta}.$$

## Third Inversion

By fixing link 3 of the slider crank mechanism, third inversion is obtained figure (3a). Now, link 2 again acts as a crank and link 4 oscillates.

Applications: Oscillating cylinder engine, crank and slotted-lever mechanism.

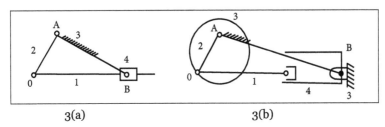

3(a)                    3(b)

Oscillating Cylinder Engine: As shown in figure (3(b)), link 4 is made in the form of a cylinder and a piston is fixed to the end of link 1. The piston reciprocates inside the cylinder pivoted to the fixed link 3. The arrangement is known as oscillating cylinder engine, in which as the piston reciprocates in the oscillating cylinder, the crank rotates.

## Fourth Inversion

If link 4 of the slider-crank mechanism is fixed, the fourth inversion is obtained figure (4(a)). Link 3 can oscillate about the fixed pivot B on link 4. This makes end A of link 2 to oscillate about B and end 0 to reciprocate along the axis of the fixed link 4.

Application: Hand Pump. In figure (4(b)) shows a hand pump. Link 4 is made in the form of a cylinder and a plunger fixed to the link 1 reciprocates in it.

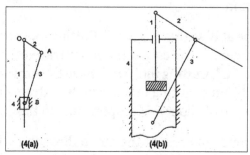

Oscillating Cylinder Engine.

## 1.3.1 Double Slider Crank Chain and their Inversion

Obtaining different mechanisms by fixing different links in a kinematic chain, is known as inversion of mechanism.

### 1. Elliptical Trammels

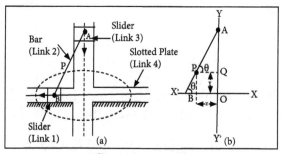

Elliptical Trammels

Let us take OX and OY as horizontal and vertical axes and let the link BA is inclined at an angle $\theta$ with the horizontal, as shown in figure (b). Now the co-ordinates of the point P on the link BA will be,

$$X = AP \cos \theta, \text{ and } Y = BP \sin \theta$$

or,

$$\frac{X}{AP} = \cos \theta; \text{ and } \frac{Y}{BP} = \sin \theta$$

Squaring and adding,

$$\frac{X^2}{(AP)^2} + \frac{Y^2}{(BP)^2} = \cos^2 \theta + \sin^2 \theta = 1$$

This is the equation of an ellipse. Hence the path traced by point P is an ellipse whose semi-major axis is AP and semi-minor axis is BP.

**Note:** If P is the mid-point of link BA, then AP = BP. The above equation can be written as,

$$\frac{X^2}{(AP)^2}+\frac{Y^2}{(AP)^2}=1 \text{ or } X^2+Y^2=(AP)^2$$

This is the equation of a circle whose radius is AP. Hence if P is the midpoint of link BA, it will tree a circle.

## 2. Oldham's Coupling

The link 1 and link 3 form turning pairs with link 2. These flanges have diametrical slots cut in their inner faces, as shown in figure (b). The intermediate piece (link 4) which is a circular disc, have two tongues (i.e., diametrical projections) $T_1$ and $T_2$ on each face at right angle to each other, as shown in figure (c). The tongues on the link 4 closely fit into the slots in the two flanges (link 1 and link 3). The link 4 can slide or reciprocate in the slots on the flanges.

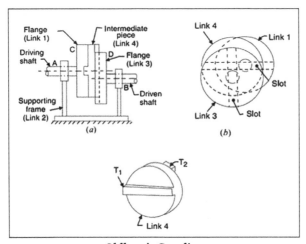

Oldham's Coupling

When the driving shaft A is rotated, the flange C (link 1) causes the intermediate piece (link 4) to rate at the same angle through which the flange has rotated and it further rotates the flange D (link 3) at the same angle and thus the shaft B rotates. Hence links 1, 3 and 4 have the same angular velocity at every instant. A little consideration will show, that there is a sliding motion between the link 4 and each of the other links 1 and 3.

If the distance between the axes of the shafts is constant, the center of intermediate piece will describe a circle of radius equal to the distance between the axes of the two shafts. Therefore, the maximum sliding speed of each tongue along its slot is equal to the peripheral velocity of the center of the disc along its circular path.

Let,

$\omega$ = Angular velocity of each shaft in rad/s.

r = Distance between the axes of the shaft in meters.

Maximum sliding speed of each tongue (in m/s), v = ω, r.

## 3. Scotch Yoke Mechanism

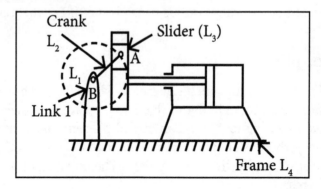

This mechanism is used for converting rotary motion into a reciprocating motion. This inversion is obtained by fixing either the link 1 or link 3.

Link 1 ⇒ fixed.

Link 2 ⇒ crank.

Link 4 ⇒ reciprocates.

## Problems

In a crank and slotted lever quick return motion mechanism, the distance between the fixed centers is 240 mm and the length of the crank is 120 mm. Let us calculate the inclination of the slotted bar with the vertical in the extreme position and the time ratio of cutting stroke to the return stroke. If the length of the slotted bar is 450 mm, let us determine the length of the stroke that passes through the extreme positions of the free end of the lever.

Solution:

Given:

$\qquad$ AC = 240 mm

$\qquad$ $CB_1$ = 120 mm

$\qquad$ $AP_1$ = 450 mm

## Inclination of the slotted bar with the vertical

Let, $\angle CAB_1$ = Inclination of the slotted bar with the vertical.

The extreme positions of the crank are shown in figure. We know that,

$$\sin \angle CAB_1 = \sin\left(90° - \frac{\alpha}{2}\right)$$

$$= \frac{B_1 C}{AC} = \frac{120}{240} = 0.5$$

$$\angle CAB_1 = 90° - \frac{\alpha}{2}$$

$$= \sin^{-1} 0.5 = 30°$$

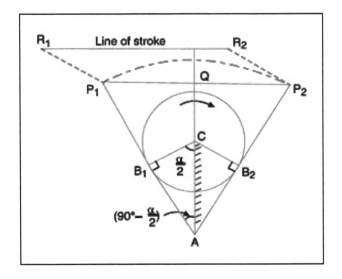

## Time ratio of cutting stroke to the return stroke

We know that,

$$90° = \frac{\alpha}{2} = 30°$$

$$\therefore \frac{\alpha}{2} = 90° - 30° = 60° \text{ or } \alpha = 2 \times 60° = 120°$$

$$\frac{\text{Time of Cutting Stroke}}{\text{Time of Return Stroke}} = \frac{360° - \alpha}{\alpha} = \frac{360° - 120°}{120°} = 2$$

## Length of the Stroke

We know that length of the stroke:

$R_1 R_2 = P_1 P_2 = 2P_1 Q = 2AP_1 \sin (90° - \alpha/2)$

$= 2 \times 450 \sin (90° - 60°) = 900 \times 0.5 = 450 \text{mm}$

## 1.4    Kinematic Analysis

### Displacement

Displacement is a vector quantity that refers to "how far out of place an object is". It is the object's overall change in position. All particles of a body move in parallel planes and travel by same distance which is known as linear displacement and is denoted by 'x'.

A body rotating about a fixed point in such a way that all particular move in circular path angular displacement and is denoted by 'θ'.

### Velocity

Rate of change of displacement is velocity. Velocity can be linear velocity or angular velocity.

Linear velocity is defined as the rate of change of linear displacement,

$$V = \frac{dx}{dt}$$

Angular velocity is defined as the rate of change of angular displacement,

$$\omega = \frac{d\theta}{dt}$$

Relation between linear velocity and angular velocity.

$$x = r\theta$$
$$\frac{dx}{dt} = r\frac{d\theta}{dt}$$
$$V = r\omega$$
$$\omega = \frac{d\theta}{dt}$$

### Acceleration

The rate of change of velocity is called as acceleration.

$f = \dfrac{dv}{dt} = \dfrac{d^2x}{dt^2}$ Linear Acceleration (Rate of change of linear velocity).

Thirdly $\alpha = \dfrac{d\omega}{dt} = \dfrac{d^2\theta}{dt^2}$ Angular Acceleration (Rate of change of angular velocity).

### Absolute Velocity

Velocity of a point with respect to a fixed point (zero velocity point).

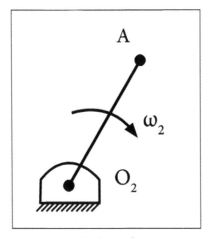

$V_a$, Absolute velocity

$$V_a = \omega_2 \times r$$
$$V_a = \omega_2 \times O_2 A$$

1. Velocity and acceleration analysis by vector polygons: Relative velocity and accelerations of particles in a common link, relative velocity and accelerations of coincident particles on separate link, Coriolis component of acceleration.

2. Velocity and acceleration analysis by complex numbers: Analysis of single slider crank mechanism and four bar mechanism by loop closure equations and complex numbers.

## Analytical Method

In function generation, the motion of input (or driver) link is correlated to the motion of output (or follower) link. Let $\theta$ and $\phi$ be the angles of rotation of input and output links respectively. Let $y = f(x)$ be the function to be generated.

The angle of rotation $\theta$ of the input link $O_2 A$ represents the independent variable x and the angle of rotation $\phi$ of the output link $O_4 B$ represents the dependent variable y, as shown in figure. The relation between x and $\theta$ and that between y and $\phi$ is generally assumed to be linear. Let $\theta_i$ and $\phi_i$ be the initial values of $\theta$ and $\phi$ representing $x_i$ and $y_i$ respectively.

$$\frac{\theta - \theta_i}{x - x_i} = r_x = \text{const.} = \frac{\theta_f - \theta_i}{x_f - x_i}$$

$$\frac{\phi - \phi_i}{y - y_i} = r_y = \text{const.} \frac{\phi_f - \phi_i}{y_f - y_i}$$

where the constants $r_x$ and $r_y$ are called scale factors. The subscripts i and f denote the initial and final values.

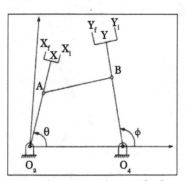

Function generation method.

## Graphical Method to Locate Precision Point

The Precision points can be obtained by the graphical method from the following steps:

- Draw a circle of radius 'b' and centre on the x-axis at a distance 'a' from point O.

- Inscribe a regular polygon of side 2n in this circle such that the two sides are perpendicular to the x-axis.

- Determine the locations of n accuracy points by projecting the vertices on x-axis as shown in figure It is sufficient to draw semi-circles only showing inscribed polygon to get the values of precision points.

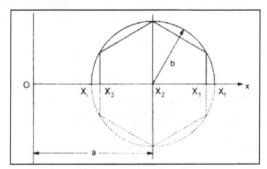

Graphical Method to Determine of Precision Point.

## Problems

1. The crank and connecting rod of a theoretical steam engine are 0.5 m and 2 m long respectively. The crank makes 180 rpm in the clockwise direction. When it has turned 45° from the inner dead centre position, let us determine:

- Velocity of piston.

- Angular velocity of connecting rod.

- Velocity of point E on the connecting rod 1.5 m from the gudgeon pin.

- Velocities of rubbing- at the pin of the crank shaft, crank and crosshead when the diameters of their pins are 50 mm, 60 mm and 30 mm respectively.

- Position and linear velocity of any point G on the connecting rod which has the least velocity relative to crank shaft.

Solution:

Given:

Crank makes = 180 rpm.

Velocity of point E on the connecting rod = 1.5 m.

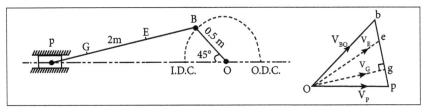

Space diagram Velocity diagram.

Vector $O_a = V_{AO} = V_A = 1.76$ m/s

$V_D$ = Vector $o_d$ = 1.6 m/s

$V_{DB}$ = Vector $b_d = 1.7$ m/s

$\omega_{BD} = V_{DB}/BD = 1.7/0.046 = 36.96$ rad/s (clockwise about B)

2. The crank of a slider crank mechanism rotates clockwise at a constant speed of 300 r.p.m. The crank is 150 min and the connecting rod is 600 mm long. Let us determine:

- Linear velocity and acceleration of the midpoint of the connecting rod.

- Angular velocity and angular acceleration of the connecting rod, at a crank angle of 45° from inner dead centre position.

Solution:

Given:

$N_{BO} = 300$rpm

$\omega_{BO} = 300/60 = 31.42$ rad/s

OB = 150 mm = 0.15m

BA = 600mm = 0.6m

We know that linear velocity of B with respect to O or velocity of B,

$$V_{BO} = V_B = \omega_{BO} \times 31.42 \times 0.15 = 4.713 \text{ m/s}$$

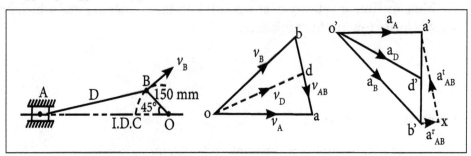

Space diagram, velocity diagram and Acceleration diagram.

Vector ob = $V_{BO} = V_B = 4.713$ m/s

$V_{AB}$ = Vector ba = 3.4 m/s

Velocity of A, $V_A$ = vector oa = 4 m/s

$V_D$ = Vector od = 4.1 m/s

$$a^r_{BO} = a_B = \frac{V^2_{BO}}{OB} = \frac{(4.713)^2}{0.15} = 148.1 \text{ m/s}^2$$

$$a^r_{BO} = \frac{V^2_{AB}}{BA} = \frac{(3.4)^2}{0.6} = 19.3 \text{m/s}^2$$

Vector c'b' = $a^r_{BO}$ = $a_B$ = 148.1 m/s²

$a_B$ = Vector o'd' = 117 m/s²

$$\omega_{AB} = \frac{V_{AB}}{BA} = \frac{3.4}{0.6} = 5.67 \text{ rad/s}^2 \text{ (Anticlockwise about B)}$$

$$\alpha_{AB} = 103 \text{ m/s}^2$$

$$\alpha_{AB} = \frac{\alpha^t_{AB}}{BA} = \frac{103}{0.6} = 171.67 \text{ rad/s}^2 \text{ (clockwise about B)}$$

3. An engine mechanism is shown in figure. The crank CB = 100 mm and the connecting rod BA = 300 mm with centre of gravity G, 100 mm from B. In the position shown, the crankshaft has a speed of 75 rad/s and an angular acceleration of 1200 rad/s². Let us determine:

- Velocity of G and angular velocity of AB.

- Acceleration of G and angular acceleration of AB.

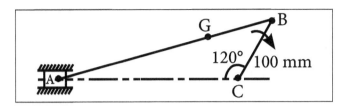

Solution:

Given:

$\omega_{BC}$= 75 red/s

$\alpha_{BC}$= 1200 rad/s²

CB= 100 mm=0.1m

BA= 300 mm=0.3 m

We know that velocity of B with respect to C, or velocity of B.

$V_{BC}=V_B=\omega_{BC} \times$ CB=75×0.1=7.5 m/s ......(perpendicular to BC)

Since the angular acceleration of the crankshaft, $\alpha_{BC}$=1200 rad/sec. therefore tangential component of we acceleration of B with roped to c.

$a^t$ BC=$\alpha_{BC} \times$ CB=1200 × 0.1 =7.5 m/s

Note: when the angular acceleration is not given, then there will be no tangential component of the acceleration.

## Velocity of G and Angular Velocity of AB

First of all draw the space diagram to some suitable scale as shown in figure (a). Now the velocity diagram shown in figure (b). is drawn as discussed below:

Draw vector cb perpendicular to CB, to some suitable scale. To represent the velocity of B with respect to C or velocity of B (ie. $V_{BC}$ or $V_B$). Such that,

Vector cb=$V_{BC}=V_B$=7.5 m/s

From point b, draw vector ba perpendicular to BA to represent the velocity of A with respect to B i.e. $V_{AB}$ and from point c, draw vector on parallel to we path of motion of A to represent the velocity of A i.e.$V_A$, The vectors ba and ca intersected at a.

Since the point G lies on AB, therefore divide vectored, ab at g in the same ratio as G divides A,B in the space diagram. In other words,

ag/ab=AG/AB

The vector cg represents the velocity of G.

By measurement, we find that velocity of G,

$$V_G = \text{vector } cg = 6.8 \text{ m/s}$$

From velocity diagram, we find that velocity of A with respect to B.

$$V_G = \text{vector } ba = 4 \text{m/s}$$

We know that angular velocity of A,B:

$$\omega_{AB} = \frac{V_{AB}}{BA} = \frac{4}{0.6} = 13.3 \text{ rad/s (Clockwise)}$$

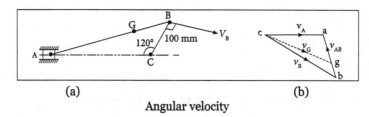

Angular velocity

## Acceleration of G and Angular Acceleration of AB

We know that radial component of the acceleration of B with respect to C.

$$a_{BC}^r = \frac{V_{AB}^2}{CB} = \frac{(7.5)^2}{0.1} = 562.5 \text{m/s}^2$$

And radial component of the acceleration of A with respect to B.

$$a_{BC}^r = \frac{V_{AB}^2}{CB} = \frac{4^2}{0.3} = 53.3 \text{m/s}^2$$

Now the acceleration diagram, as shown in figure is drawn as discussed below.

Draw vector c'b'' parallel to CB to some suitable scale, to represent component of the acceleration of B with respect to C i.e $a^r$ BC such that Vector b''b' = $a^r$ BC=562.5 m/s$^2$.

From point b'' draw vector b''b' perpendicular to vector c'b'' or CB represent the tangential component of the acceleration of B with respect to i.e $a^r$ BC such that,

$$\text{Vector b''b' = } a^r \text{ BC = 120 m/s}^2$$

Join c'b' the vector c'b' represents the total acceleration of B with respect to C i.e. $a_{BC}$.

From point b'x parallel to BA to represent radial component of the acceleration of A with respect to B i.e $a^r$ AB such that from point x, draw vector xa' perpendicular to vector b'x or BA to represent tangential component of the acceleration of, A with respect to B i.e. a$^t$AB, whose magnitude is not yet known.

Now draw vector c'a' parallel to the path of motion of A (which is along A C) to represent the acceleration of A i.e. $a_A$, The vectors xa' and c'a' intersect at a'. Join b'a' The vector b'a' represents the acceleration of A with respect to B i.e. $a_{AB}$.

In order to find the acceleration of G. Divide vector a'b' in g' in the same ratio as G divides BA in figure. Join c'g'. The vector c'g' represents the acceleration on of G.

By Measurement we find that acceleration of G.

$a_G$=vector c'g'=414 m/s²

From acceleration diagram we find that tangential component of the acceleration of A with respect to B.

## 1.4.1 Instantaneous Centre Method

Let us consider a plane body P having a nonlinear motion relative to another body q consider two points A and B on body P. Let the velocities be $V_a$ and $V_b$ respectively.

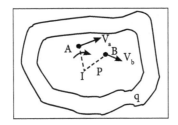

If a line is drawn perpendicular to $V_a$, at A the body can be imagined to rotate about some point on the line. Thirdly, centre of rotation of the body also lies on a line perpendicular to the direction of $V_b$ at B. If the intersection of the two lines is at I, the body P will be rotating about I at that instant. The point I is known as the instantaneous centre of rotation for the body P. The position of instantaneous centre changes with the motion of the body.

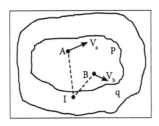

In case of the perpendicular lines drawn from A and B meet outside the body P as shown in the above figure.

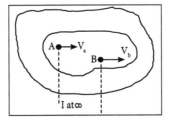

If the direction of $V_a$ and $V_b$ are parallel to the perpendicular at A and B met at ∞. This is the case when the body has linear motion.

## Number of Instantaneous Centers

The number of instantaneous centers in a mechanism depends upon number of links N is the number of instantaneous centers and n is the number of links.

$$N = \frac{n(n-1)}{2}$$

## Type of Instantaneous Centers

There are three types of instantaneous centers namely fixed, permanent and neither fixed nor permanent.

Example: ram bar mechanism n=4,

$$N = \frac{n(n-1)}{2} = \frac{4(4-1)}{2} = 6$$

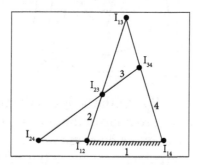

Fixed instantaneous center $I_{12}$, $I_{14}$.

Permanent instantaneous center $I_{23}$, $I_{34}$.

Neither fixed nor permanent instantaneous center $I_{13}$, $I_{24}$.

## 1.4.2 Relative Velocity Method
### Relative Velocity Equation

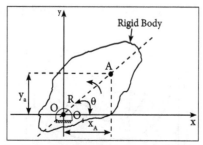

Point O fixed and A is a point on rigid body.

Rotation of a rigid link about a fixed center.

Consider rigid link rotating about a fixed center O, as shown in figure. The distance between o and A is R and OA makes and angle θ with x-axis next link.

$$x_A = R\cos\theta, \ y_A = R\sin\theta$$

Differentiating $x_A$ with respect to time gives velocity,

$$\frac{dx_A}{dt} = R(-\sin\theta)\frac{d\theta}{dt}$$
$$= -R\omega\sin\theta$$

Similarly,

$$\frac{dy_A}{dt} = R(-\cos\theta)\frac{d\theta}{dt}$$
$$= -R\omega\cos\theta$$

Let,

$$\frac{dx_A}{dt} = V_A^x \qquad \frac{dy_A}{dt} = V_A^y$$

$$\omega = \frac{d\theta}{dt} = \text{angular velocity of OA}$$
$$\therefore V_A^x = -R\omega\sin\theta$$
$$V_A^x = -R\omega\cos\theta$$

Total velocity of point A is given by,

$$V_A = \sqrt{(-R\omega\sin\theta)^2 + (-R\omega\cos\theta)^2}$$
$$V_A = R\omega$$

## Relative Velocity Equation of Two Points on a Rigid link

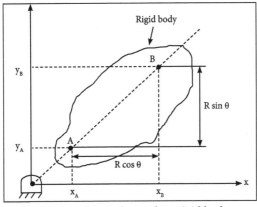

Point A and B are located on rigid body.

From figure,

$$x_B = x_A + R\cos\theta \quad y_B = y_A + R\sin\theta$$

Differentiating $x_B$ and $y_B$ with respect,

$$\frac{dx_B}{dt} = V_B^x = \frac{dx_A}{dt} + R(-\sin\theta)\frac{d\theta}{dt}$$

$$= \frac{dx_A}{dt} + R\omega\sin\theta = V_A^x - R\omega\sin\theta$$

Similarly,
$$\frac{dy_B}{dt} = V_B^y = \frac{dy_A}{dt} + R(\cos\theta)\frac{d\theta}{dt}$$

$$= \frac{dy_A}{dt} + R\omega\cos\theta = V_A^y - R\omega\sin\theta$$

$$V_A \; V_A^x \leftrightarrow V_A^y = \text{Total velocity of point A}$$

Similarly, $V_B \; V_B^x \leftrightarrow V_B^y = \text{Total velocity of point B}$

$$= V_A^x \leftrightarrow (R\omega\sin\theta) \leftrightarrow V_A^y \leftrightarrow R\omega\cos\theta$$

$$= (V_A^x \leftrightarrow V_A^y) \leftrightarrow (R\omega\sin\theta + R\omega\cos\theta)$$

$$= (V_A^x \leftrightarrow V_A^y)\,VA \text{ Similarly, } (R\omega\sin\theta + R\omega\cos\theta) = R\omega$$

$$\therefore V_B = V_A \leftrightarrow R\omega = V_A \leftrightarrow V_B A$$
$$\therefore V_{BA} = V_B - V_A$$

## Problems

The following data refer to the dimensions of the links of a four-bar mechanism: AB = 50 mm, BC = 66 mm, CD = 56 mm and AD (fixed link) = 100 mm. At the instant when $\angle DAB = 60°$, the link AB has an angular velocity of 10.5 rad/s in the counter clockwise direction. Let us determine the Velocity of Point C, Velocity of Point E on the link BC while BE = 40 mm and the angular velocities of the links BC and CD and also sketch the mechanism and Indicate the data.

Solution:

Given:

AB = 50 mm

BC = 66 mm

CD = 56 mm

AD = 100 mm

$\angle DAB = 60°; \omega_{BA} = 10.5 \, \text{rad/sec}$

$V_{BA} = \omega_{BA} \times AB = 10.5 \times 0.05 = 0.525 \text{ m/s.}$

# a. Configuration Diagram

Scale: 10 mm = 1 cm

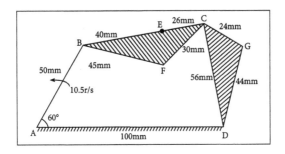

# b. Velocity Diagram

Scale: 0.105 m/s = 1 cm

Directly from the velocity diagram:

$V_{BA}$ = ab = 0.525 m/s (already known)

$V_{FA}$ = af = 0.504 m/s

$V_{GA}$ = ag = 0.315 m/s

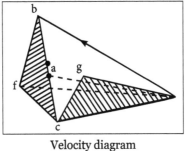

Velocity diagram

## Procedure

## a. Configuration Diagram

Refer figure above:

- As clearly data given, we can draw ABCD.

- From point 'B' draw an arc of 45 mm radius and from C draw an arc of 30 mm radius then the arc intersecting point will be 'F'.

- From 'D', draw an arc of 44 mm radius and from C draw an arc of 24 mm radius then the arc intersecting point will be 'G'.

## Velocity Diagram

Refer Figure Above

- Draw ab = 0.525 m/s with anticlockwise direction at any point and perpendicular to AB.

- Since $\overline{AD}$ is fixed, the relative velocity about these points is zero, hence a and b, d will be on the same point.

- Draw a line perpendicular to $\overline{CD}$ at 'd' and draw a line perpendicular to $\overline{BF}$ at 'b' to cut each other.

Locate the cutting point as 'c'.

- To locate the offset point F: Draw a line perpendicular to $\overline{BF}$ at 'b' and line perpendicular to $\overline{CF}$ at 'c' and locate the cutting point as 'f' and joint $\overline{af}$.

- To locate the offset point G: Draw a line perpendicular to $\overline{CV}$ at 'c' and line perpendicular to $\overline{GD}$ at 'd' and locate the cutting point as 'g'.

To locate 'E' on the velocity diagram:

$$\frac{V_{be}}{V_{bc}} = \frac{BE}{BC}$$

$$V_{be} = V_{bc} \times \frac{BE}{BC} = 0.34 \times \frac{40}{66} = 0.206 \, \text{m/s.}$$

## Results

i. Velocity of point 'C' with respect to 'a':

$$V_{ca} = 3.75 \times 0.105 = 0.393 \, \text{m/s}$$

$$V_{ca} = \text{Length 'ca'} \times \text{Actual scale}$$

ii. Velocity of point E with respect to B:

$$V_{EB} = 0.206 \, \text{m/s}$$

## iii. Angular velocities of link BC and CD:

$$\omega_{BC} = \frac{V_{BC}}{BC} \left[\because V_{BC} = 3.25 \times 0.105; \ V_{BC} = 0.341\,\text{m/s}\right]$$

$$= \frac{0.341}{0.066} = 5.17\,\text{rad/sec}$$

$$\omega_{CD} = \frac{V_{CD}}{CD} = \frac{0.393}{0.056} \left[\because V_{CD} = 3.75 \times 0.105 = 0.393\,\text{m/s}\right]$$

$$= \frac{0.393}{0.056} = 7.01\,\text{rad/sec}$$

## iv. Velocity of offset, point F with respect to 'a':

$V_{FA} = 0.504$ m/s (Direct measurement)

## v. Velocity of offset point G with respect to 'a':

$V_{GA} = 0.315$ m/s (Direct measurement)

## vi. To calculate Rubbing Velocities:

Velocity of rubbing at pin B = $V_B = (\omega_{AB} + \omega_{BC}) \times r_B$

$= (10.5 + 5.17) \times 0.040$

$V_B = 0.6268$ m/s

Velocity of rubbing at pin C = $V_c = (\omega_{CD} + \omega_{BC})\, r_C$

$= (7.01 + 5.17) \times 0.025$

$V_c = 0.3045$ m/s

Velocity of rubbing at pin A = $V_A = (\omega_{AD} + \omega_{AB})\, r_A$

$= (10.5) \times 0.03$

[ $\omega_{AD} = 0$; sine fixed]

$V_A = 0.315$ m/s

Velocity of rubbing at pin 'D' = $V_D = (\omega_{AD} + \omega_{CD}) \times r_D$

$= (7.01) \times 0.035$

$[\omega_{AD} = 0]$

$V_D = 0.2454$ m/s.

## 1.4.3 Kennedy Theorem

## Arnold Kennedy Theorem of Three Centers

It states that "If three bodies have motion relative to each other, their instantaneous centers should lie in a straight line".

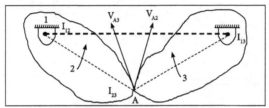

Elative motion between two bodies.

Consider a three link mechanism with link 1 being fixed link 2 rotating about $I_{12}$ and link 3 rotating about $I_{13}$. Therefore $I_{12}$ and $I_{13}$ are the instantaneous centers for link 2 and link 3. Let us assume that instantaneous center of link 2 and 3 be at point A i.e. $I_{23}$. Point A is a coincident point on link 2 and link 3.

Considering A on link 2, velocity of A with respect to $I_{12}$ will be a vector $V_{A2}$ r to link A $I_{12}$. Similarly for point A on link 3, velocity of A with respect to $I_{13}$ will be r to A $I_{13}$. It is seen that velocity vector of $V_{A2}$ and $V_{A3}$ are in different directions which is impossible. Hence, the instantaneous center of the two links cannot be at the assumed position.

It can be seen that when $I_{23}$ lies on the line joining $I_{12}$ and $I_{13}$ the $V_{A2}$ and $V_{A3}$ will be same in magnitude and direction. Hence, for the three links to be in relative motion all the three centers should lie in a same straight line.

## Steps to Locate Instantaneous Centers

Step 1: Draw the configuration diagram.

Step 2: Identify the number of instantaneous centers by using the relation.

$$N=\frac{(n-1)n}{2}$$

Step 3: Identify the instantaneous centers by circle diagram.

Step 4: Locate all the instantaneous centers by making use of Kennedy's theorem.

## Problems

A slider crank mechanism has lengths of crank and connecting rod equal to 200 mm and 200 mm respectively locate all the instantaneous centers of the mechanism for the position of the crank when it has turned through 30° from IOC. Let us determine the velocity of slider and angular velocity of connecting rod if crank rotates at 40 rad/sec.

Solution:

Given:

Step 1: Draw configuration diagram to a suitable scale.

Step 2: Determine the number of links in the mechanism and find number of instantaneous centers.

$$N = \frac{(n-1)n}{2}$$

$$n = 4 \text{ links} \qquad N = \frac{4(4-1)}{2} = 6$$

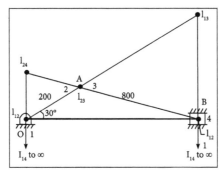

Configuration Diagram

Step 3: Identify instantaneous centers.

Suit it is a 4-bar link the resulting figure will be a square.

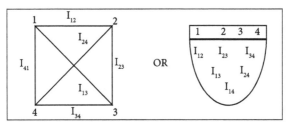

Locate fixed and permanent instantaneous centers. To locate neither fixed nor permanent instantaneous centers use Kennedy's three centers theorem.

Step 4: Velocity of different points.

$$V_a = \omega_2 AI_{12} = 40 \times 0.2 = 8 \text{m/s}$$

$$\text{Also } V_a = \omega_2 \times A_{13}$$

$$\omega_3 = V_a / AI_{13}$$

$$V_b = \omega_3 \times BI_{13} = \text{Velocity of slider}$$

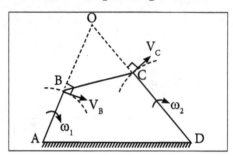

## 1.5    Velocity in Four Bar Mechanism

The figure shows a four bar mechanism, which consists of a fixed link AD, two movable links AB and CD rotating about points A and D respectively and a connecting link BC (which is also known as coupler BC). Let the link AB is rotating at a uniform angular velocity and it is required to find the corresponding motions of the two links BC and CD.

Let,

$\omega_{CD}$ = Angular velocity of link AB, rotating about A in the clock-wise direction. The value of $\omega_{AB}$ is given as $\omega_1$.

$\omega_{CD}$ = Angular velocity of link CD, rotating about D. The value of $\omega_{CD}$ is to be calculated which is $\omega_2$.

$\omega_{BC}$ = Angular velocity of the link $\omega_{BC}$. The value of cow is also to be calculated.

$V_B$ =Linear velocity of point B in the direction perpendicular to AB.

$$= \omega_{AB} \times AB \qquad \qquad ...(1)$$

$V_c$ =Linear velocity of point C in the direction perpendicular to CD.

$$= \omega_{CD} \times CD \qquad \qquad ...(2)$$

The link AB and link CD are having motion of rotation, whereas the link BC is having motion of translation as well as rotation. The instantaneous centre of link BC for the given position is determined by drawing normals to the directions of velocity $V_B$ and $V_C$ The normal to the direction $V_B$ a is the line AB whereas the normal to the direction Vc is line CD.

Hence produce line AB and CD. The intersection of these lines gives the point o (i.e., instantaneous centre) for the link BC.

The linear velocity at B is given by,

$$V_B = \omega_{BC} \times BO \qquad \qquad ...(3)$$

and linear velocity at C is given by,

$$V_c = \omega_{BC} \times CO \qquad \qquad ...(4)$$

Equating equations (1) and (3), we get,

$$\omega_{AB} \times AB = \omega_{BC} \times BO$$

$$\omega_{BC} = \frac{\omega_{BC} \times AB}{BO}$$

In equation the values $\omega_{AB}$ are given. The value of BO is obtained from figure. Hence, $\omega_{BC}$(angular velocity of link BC) can be determined. Again equating equations (2) and (4), we get,

$$\omega_{CD} \times CD = \omega_{BC} \times CO$$

$$\omega_{CD} = \frac{\omega_{BC} \times CO}{CD}$$

In the above equation value of CD is obtained from figure and value of $\omega_{ac}$ and $\omega_{CD}$.

## 1.5.1 Slider Crank Mechanism

In a crank and slotted lever mechanism crank rotates at 300 rpm in a counter clockwise direction.

- Angular velocity of connecting rod.

- Velocity of slider.

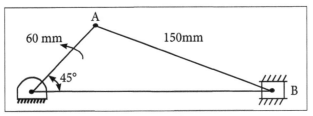

Configuration Diagram

Step 1: Determine the magnitude and velocity of point A with respect to o.

Step 2: Choose a suitable scale to draw velocity vector diagram.

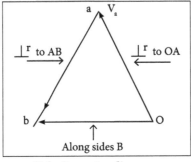

Velocity vector diagram.

$$V_{ab} = \overrightarrow{ab} = 1300\,\text{mm/sec}$$

$$\omega_{ba} = \frac{V_{ba}}{BA} = \frac{1300}{150} = 8.66\,\text{rad/sec}$$

$$V_b = \overrightarrow{ob}\,\text{velocity of slider}$$

Note: Velocity of slider is along the line of sliding.

## 1.5.2 Rubbing Velocity at a Pin-Joint

The two links 1 and 2 are connected by means of a pin-joint as shown in figure.

Let,

$\omega_1$=Angular velocity of link 1,

$\omega_2$= Angular velocity of link 2,

r = Radius of the pin at the joint.

The rubbing velocity is defined as the algebraic difference between the angular velocities of the two links which are connected by pin-joints, multiplied by the radius of the pin.

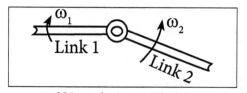

Rubbing velocity at a Pin-joint.

Hence the rubbing velocity at the pin joint, when the two connected links move in opposite direction is given by.

Rubbing velocity $= r(\omega_1 + \omega_2)$

But if the two connected links move in the same direction, then,

Rubbing velocity $= (\omega_1 - \omega_2)$

If a pin connects one sliding member and the other turning member (For example gudgeon pin of a connecting rod) the angular velocity of the sliding member is zero and hence the velocity of rubbing will be given by,

Rubbing velocity $= r \times \omega$

Where,

r= Radius of the pin.

$\omega$= Angular velocity of the turning member.

## Problems

1. The angular velocity of the crank OA is 600 r.p.m. Let us determine the linear velocity of slider 6 and angular velocity of the link 5.

Solution:

Given:

$N_{OA}$=600 r.p.m., $\theta$=75°

OA=28 mm

AB=44 mm

BC=49 mm

BD=46 mm

OC=65 mm

$$\omega_{OA} = \frac{2\pi N_{OA}}{60} = \frac{2\pi \times 600}{60} = 20\pi \, \text{rad/sec}$$

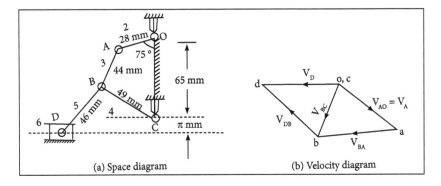

(a) Space diagram      (b) Velocity diagram

Velocity of A with respect O,

$$V_{AO} = \omega_{OA} \times OA = 20\pi \times 28 \, \text{mm/s}$$
$$= \frac{20\pi \times 28}{1000} \text{m/s}$$
$$= 1.75 \, \text{m/s}$$

The point O and C are fixed in the given mechanism. Hence these points will be located at a common point in the velocity diagram.

As point O is fixed, hence the velocity of A with respect to O is also equal to velocity of A,

$$V_{AO} = V_A = 1.758 \, \text{m/s}$$

First draw the space diagram to a suitable scale as shown in figure. Now draw the velocity diagram as shown in Figure (b) as per method given below:

Take any point o. Here fixed point should be taken. Choose a suitable velocity scale and draw vector oa perpendicular to OA to represent the velocity of A with respect to o. Cut oa = 1.758 m/s.

The velocity of B with respect to A is perpendicular to AB whereas the velocity of B with respect to C is perpendicular to BC. From points draw vector ob perpendicular to AB to represent the velocity of B with respect to A But the magnitude of am is not known.

Since OC is fixed. Hence points o and C in velocity diagram will coincide. Now from point. Draw $B_{cb}$ (or ob) perpendicular to BC so as to intersect vector ab in b. Now vector ob represent the velocity of B with respect to C.

The velocity of point D with respect to point B is perpendicular to BD. Also the point D is moving in a horizontal direction. From point b, draw vector bd perpendicular to BD to represent the velocity of D with respect to B. The magnitude of $v_{DB}$ is not known.

The velocity of the slider 6 or of point O is in the horizontal direction. Hence draw vectored parallel to the path of motion of the slider 6 to represent the velocity of D. The vectors bd and ad intersect at d. This completes the velocity diagram.

By measurement from velocity diagram, we get,

$\qquad v_D$=vector od=6.6 m/s

$\qquad v_{DB}$=Vector bd=7 m/s

i. Linear velocity of slide 6.

$\qquad$ The linear velocity of slider 6 is equal to $v_D$.

$\qquad$ Linear velocity of slider 6=vD=6.6 m/s.

ii. Angular velocity of link 5 or link BD.

$\qquad v_{DB}$=Vector bd=7 m/s

$\qquad v_{DB}$ = Vector bd=7 m/s

$\qquad v_{DB} \times$ BD

$$\omega_{DB} = \frac{v_{DB}}{BD} = \frac{7}{\left(\dfrac{46}{1000}\right)}$$

$$= \frac{7 \times 1000}{46}$$

$$= 152.17 \text{ rad/s}$$

## Four – Bar Mechanism

2. In a four bar chain ABCD link AD is fixed and 15 cm long. The crank AB is 4 cm long rotates at 180 rpm (cw) while link CD rotates about D is 8 cm long and BC = AD and BAD = 60°. Let us determine the angular velocity of link CD.

Solution:

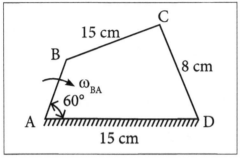

Configuration Diagram

## Velocity Vector Diagram

$$V_b = \omega r = \omega_{ba} \times AB = \frac{2\pi \times 120}{60} \times 4 = 50.24 \, cm/sec$$

## Choose a Suitable Scale

$$1cm = 20 \, m/s = \overrightarrow{ab}$$

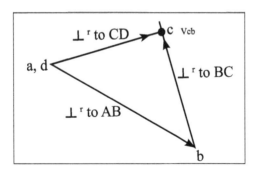

$$V_{cb} = \overrightarrow{bc}$$
$$V_c = \overrightarrow{dc} = 38 \, cm/sec = V_{cd}$$

We know that, $V = \omega R$

$$V_{cb} = \omega_{CD} \times CD$$

$$\omega_{CD} = \frac{V_{cd}}{CD} = \frac{38}{8} = 4.75 \, rad/sec \, (cw)$$

## Corioli's component of acceleration and its application

### Coriolis Acceleration

A slider attached to ground experiences only sliding acceleration. A slider attached to a rotating link (such that the slider is moving in or out along the link as the link rotates) experiences all 4 components of acceleration.

Perhaps the most confusing of these is the Coriolis acceleration, though the concept of Coriolis acceleration is fairly simple. Imagine ourself standing at the center of a merry-go-round as it spins at a constant speed. We begin to walk toward the outer edge of the merry-go-round at a constant speed ($d_r/d_t$).

Even though we are walking at a constant speed and the merry-go-round is spinning at a constant speed, our total velocity is increasing because we are moving away from the center of rotation. This is the coriolis acceleration.

Coriolis component of acceleration comes to picture when a slider is sliding along a rotating link. The tangential component of acceleration of slider with respect to the coincident point on the link is known as Coriolis component of acceleration.

The expression for Coriolis component of acceleration is:

$$f_{Cr} = 2\,V\omega$$

where,

$f_{Cr}$ = Coriolis acceleration of the particle.

V = Velocity of the slider.

$\omega$ = Angular velocity of the link.

- Whit worth quick-return mechanism.

- Crank and slotted lever quick return motion mechanism.

### Problems

1. Let us locate the instantaneous center's of the slider crank mechanism shown in the figure. Let us also determine the velocity of the slider.

Solution:

Given:

- $\omega_{OB}$ = 12 rad/s

- OB = 100 mm = 0.1 m

We know that linear velocity of the crank OB,

$$V_{OB} = V_B = \omega_{OB} \times OA$$

$$= 12 \times 0.10 = 1.2 \text{ m\textbackslash s}$$

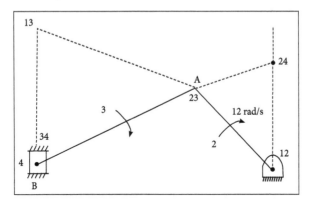

## Location of Instantaneous Centers

$$N = \frac{n \cdot (n-1)}{2} = \frac{4(4-1)}{2}$$

$$N = 6$$

The instantaneous centers in a slider frank mechanism are located as discussed below:

i. Since there are four links (i.e., n = 4), therefore the number of instantaneous centers:

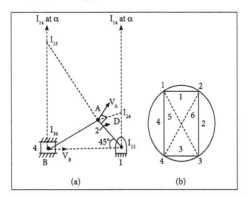

(a)   (b)

ii. Locate the fixed and permanent instantaneous centers by inspection. These centers are $I_{12}$, $I_{23}$ and $I_{34}$ as shown in the figure (a). Since the slides (link 4) moves on a straight surface (link 1), therefore the instantaneous center $I_{14}$ will lie at infinite.

iii. Locate the other two remaining neither fixed nor permanent instantaneous centers, by Kennedy's theorem. This is done by circle diagram as shown in figure (b). Mark four points 1, 2, 3 and 4 (equal to the number of links in a mechanism) on the circle to indicate $I_{12}$, $I_{23}$, $I_{34}$ and $I_{14}$.

iv. Join 1 to 3 to form two triangles 123 and 341 in the circle diagram. The side 13, common to both triangles, is responsible for completing the two triangles. Therefore the center $I_{13}$ will lie on the intersection of $I_{12} I_{23}$ and $I_{14} I_{34}$, produced if necessary. Thus center $I_{13}$ is located. Join 1 to 3 by a dotted line and mark number 5 on it.

v. Join 2 to 4 by a dotted line to form two triangles 234 and 124. The side 24, common to both triangles, is responsible for completing the two triangles. Therefore the center $I_{24}$ lies on the intersection of $I_{23} I_{34}$ and $I_{12} I_{14}$. Join 2 to 4 by a dotted line on the circle diagram and mark number 6 on it.

Thus all the six instantaneous centers are located:

By measurement we find, that,

$$I_{13} A = 560 \text{ mm} = 0.56 \text{ m},$$

$$I_{13} B = 460 \text{ mm} = 0.46 \text{ m},$$

Velocity of the slider A.

Let, $V_A$ = Velocity of the slider A.

We know that,

$$\frac{V_A}{I_{13}A} = \frac{V_B}{I_{13}B}$$

$$V_B = V_A \times \frac{I_{13}B}{I_{13}A}$$

$$V_B = 1.2 \times \frac{0.46}{0.56}$$

VB = 0.98 m/s.

2. A slider crank mechanism has lengths of crank and connecting rod equal to 200 mm and 200 mm respectively. Let us locate all the instantaneous centers of the mechanism for the position of the crank when it has turned through 30° from IOC. Let us also determine the velocity of slider and angular velocity of connecting rod if crank rotates at 40 rad/sec.

Solution:

Given:

Step 1: Draw configuration diagram to a suitable scale.

Step 2: Determine the number of links in the mechanism and find number of instantaneous centers.

$$N = \frac{(n-1)n}{2}$$

$n = 4$ links       $N = \frac{4(4-1)}{2} = 6$

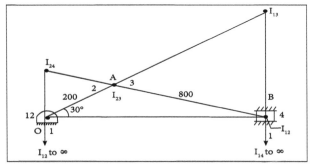

Configuration Diagram

Step 3: Identify instantaneous centers. Suit it is a 4-bar link the resulting figure will be a square.

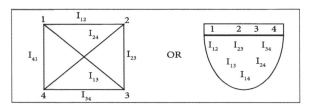

Locate fixed and permanent instantaneous centers. To locate neither fixed nor permanent instantaneous centers use Kennedy's three centers theorem.

Step 4: Velocity of different points,

$V_a = \omega_{2A} \, I_{12} = 40 \times 0.2 = 8$ m/s

Also $V_a = \omega_2 \times A_{13}$

$\omega_3 = V_a / AI_{13}$

$V_b = \omega_3 \times BI_{13} =$ Velocity of slider

3. A slider sliding at 100 mm/sec on a link, which is rotating at 6 rpm is subjected to Coriolis acceleration. Let us find its magnitude.

Solution:

Given:

$V = 100$ mm/s

$N = 60$ rpm

Coriolis acceleration:

$$\omega = \frac{2\pi N}{60} = \frac{2\pi(60)}{60} = 6.28\,\text{rad/s}$$

$$= 2 \times 6.28 \times 100$$

$$a_o = 1{,}256\,\text{mm/s}.$$

# Inertia Forces in Reciprocating Parts

## 2.1 Reciprocating Engine

### Velocity and Acceleration of Piston in a Reciprocating Engine

The various forces acting on the reciprocating parts of a horizontal engine. Consider the motion of a crank and connecting rod of a reciprocating stream engine. Let OC be the crank and PC the angular velocity of $\omega$ rod/s and the crank turns through an angle $\theta$ from the IDC.

Let,

x = Displacement of a reciprocating body.

P = From I.D.C after time t seconds, during which the crank has turned through an angle $\theta$.

l = Length of connecting rod between the centers.

r = Radius of crank or crank pin circle.

$\phi$ = Inclination of C.V to the line of stroke PO.

n = Ratio of length of connecting rod to the radius of crank = $1/\gamma$.

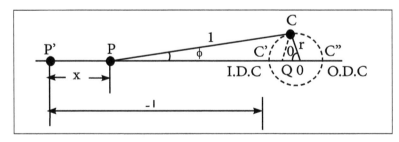

### Velocity of the Piston

From the geometry,

$$x = P'P = OP' - OP = (P'C' + C'O) - (PQ + QO)$$
$$= (1 + \gamma) - (1 \cos \phi + \gamma \cos \theta)$$

$$= \gamma(1-\cos\theta)+1(1-\cos\phi)$$

$$= \gamma\left[(1-\cos\theta)+\frac{1}{\gamma}(1-\cos\phi)\right]$$

$$= \gamma\left[(1-\cos\theta)+n(1-\cos\phi)\right] \qquad ...(1)$$

From $\triangle$ CPQ and CQO,

$$CQ=1\sin\phi=\gamma\sin\theta(or\,)\frac{1}{\gamma}=\frac{\sin\theta}{\sin\phi}$$

$$n=\sin\theta/\sin\phi(or)\ \sin\phi=\sin\theta/n \qquad ...(2)$$

$$\cos\phi=\left(1-\sin^2\phi\right)^{1/2}=\left(1-\frac{\sin^2\theta}{n^2}\right)^{1/2}$$

Expanding the above expanding by binomial theorem, we get,

$$\cos\phi=1-\frac{1}{2}\times\frac{\sin^2\theta}{n^2}+...$$

$$1-\cos\phi=\frac{\sin^2\theta}{2n^2} \qquad ...(3)$$

Substituting the value of $(1-\cos\phi)$ in equation (1); we have,

$$x=r\left[(1-\cos\theta)+n\times\frac{\sin^2\theta}{2n^2}\right]$$

$$=r\left[(1-\cos\theta)+\frac{\sin^2\theta}{2n}\right] \qquad ...(4)$$

Differentiating equation (4) with respect to $\theta$,

$$\frac{dx}{d\theta}=r\left[\sin\theta+\frac{1}{2n}\times2\sin\theta\cdot\cos\theta\right]$$

$$=r\left(\sin\theta+\frac{\sin 2\theta}{2n}\right) \qquad ...(5)$$

Velocity of P with respect to O (or) Velocity of the piston P,

$$\gamma_{PO}=\gamma_P=\frac{dx}{dt}=\frac{dx}{d\theta}\times\frac{d\theta}{dt}=\frac{dx}{d\theta}\times\omega$$

Substituting the value of $\dfrac{dx}{d\theta}$ from equation (5),

$$\gamma_{PO} = \gamma_P = \omega \cdot \gamma \left( \sin \theta + \frac{\sin 2\theta}{2n} \right)$$

W.K.T by Klein's construction,

$$\gamma_P = \omega \times OM$$

By comparing this equation, we find that,

$$OM = r \left( \sin \theta + \frac{\sin 2\theta}{2M} \right) \qquad \qquad ...(6)$$

## Acceleration of the Piston

Since the acceleration is the rate of change of velocity, therefore acceleration of the piston P,

$$a_P = \frac{d\gamma_P}{dt} = \frac{d\gamma_P}{d\theta} \times \frac{d\theta}{dt} = \frac{d\gamma_P}{d\theta} \times \omega$$

Differentiating equation (6) with respect to $\theta$,

$$\frac{d\gamma_P}{d\theta} = \omega \cdot r \left[ \cos \theta + \frac{\cos 2\theta \times 2}{2n} \right]$$

$$= \omega \cdot r \left[ \cos \theta + \frac{\cos 2\theta}{n} \right]$$

Substituting value of $\dfrac{d\gamma_P}{d\theta}$ in above equations, we have,

$$a_P = \omega \cdot r \left[ \cos \theta + \frac{\cos 2\theta}{n} \right] \times \omega$$

$$a_P = \omega^2 \cdot r \left[ \cos \theta + \frac{\cos 2\theta}{n} \right] \qquad \qquad ...(7)$$

## Reciprocating Steam Engine Mechanism

Consider the slider crank mechanism shown in the Figure. OC is the crank, BC the connecting rod, (the crank pin and B the gudgeon pin on the cross head and P is the piston. The crank is rotating clockwise with angular speed $\omega$.

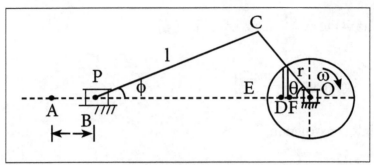

Slider Crank Mechanism.

Velocity and acceleration of the piston: Displacement of the piston from top dead centre:

$$x = AB = EF$$
$$= ED + DF$$
$$= (OE - OD) + (BF - BD)$$
$$= (r - r \cos \theta) + (l - l \cos \phi)$$
$$= r(1 - \cos \theta) + l(1 - \cos \phi).$$

Now, $CD = r \sin \theta = l \sin \phi$

Or, $\sin \phi = \left( \dfrac{r}{l} \sin \theta \right) = \dfrac{\sin \theta}{n}$

Where, $\dfrac{l}{r} = n$ is the ratio of length of connecting rod to that of crank.

$$\cos \phi = \left[ 1 - \sin^2 \phi \right] 0.5$$
$$= \dfrac{\left( n^2 - \sin^2 \theta \right)^{0.5}}{n}$$

Therefore,

$$n = r(1 - \cos \theta) + l \left[ 1 - \dfrac{\left( n^2 - \sin^2 \theta \right)^{0.5}}{n} \right] \qquad \dots(1)$$

$$= r \left[ (1 - \cos \theta) + n - \left( n^2 - \sin^2 \theta \right)^{0.5} \right]$$

Velocity of piston:

$$V_P = \frac{dx}{dt}$$

$$V_P = \frac{dx}{d\theta} \cdot \frac{d\theta}{dt}$$

$$= \omega \cdot \frac{dx}{d\theta}$$

$$= \omega_r \left( \sin\theta + \left(\frac{1}{2}\right)(n_2 - \sin^2\theta)^{-0.5} \, 2\sin\theta\cos\theta \right)$$

$$= \omega_r \left( \sin\theta + \frac{\sin^2\theta}{2(n^2 - \sin^2\theta)^{-0.5}} \right)$$

$$= \omega_r \left( \sin\theta + \frac{\sin^2\theta}{2n} \right) \qquad \qquad ...(2)$$

Acceleration of piston,

$$f_p = \frac{dv}{dt}$$

$$f_p = \frac{dv}{d\theta} \cdot \frac{d\theta}{dt}$$

$$= \omega \frac{dv}{d\theta}$$

$$= \omega^2 r \left[ \frac{-(n^2 - \sin^2\theta)^{0.5} \cdot 2\cos^2\theta}{(n^2 - \sin^2\theta)} \right]$$

$$= \omega^2 r \left[ \cos\theta - \frac{0.5\sin^2\theta + \cos^2\theta(n^2 - \sin^2\theta)}{(n^2 - \sin^2\theta)} \right]$$

$$f_p = \omega^2 r \left[ \cos\theta + \frac{\cos^2\theta}{n} \right]$$

## 2.1.1 Angular Velocity and Angular Acceleration of Connecting Rod Analytical Method

Consider the Slid crank chain, as shown in figure.

Let,

$W_c$ = weight of the connecting rod.

$M_c$ =Mass of the reciprocating part.

K = Radius of gyration of connecting rod about its center of gravity and perpendicular to the plane of rotation.

L = Distance of Centre of gravity G of connecting rod from the gudgeon pin.

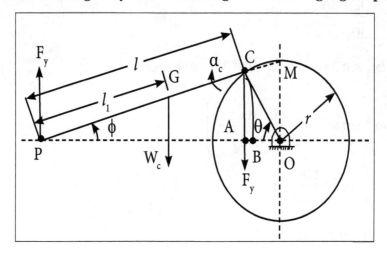

l=length of connecting rod,

$$L = l_1 + \frac{K^2}{l_1}$$

r = Radius of crank.

Total equivalent reciprocating mass, $M_{re} = M_r + (1 - l_1)\frac{W_c}{gl}$.

The inertia force due to, $M_{re}$,

$$F_i = -M_{re} \times f_p$$

Where,

$f_p$ = Acceleration of reciprocating part,

$$= \omega^2 r \left( \cos\theta + \frac{\cos 2\theta}{n} \right)$$

$$n = \frac{l}{r}$$

Torque exerted on the crankshaft due to inertia force,

$$T_i = F_i \times OM$$

Where, $OM = \dfrac{r\sin(\theta + \phi)}{\cos\phi}$ and $\sin\phi = \dfrac{\sin\theta}{n}$

Correction couple, $T_o = W_c l_1 \dfrac{(1-L)\alpha_c}{g}$

Where,

$\alpha_c$ = Angular acceleration of connecting rod,

$$= -\omega^2 \dfrac{\left(n^2 - 1\sin\theta\right)}{\left(n^2 - \sin^2 2\theta\right)^{1.5}}$$

$$= -\omega^2 \dfrac{\sin\theta}{n}$$

Let $F_y$ be two equal and opposite forces applied at P and C.

Then,

$$F_y \cdot AP = T_o \dfrac{W_c l_1 (1-L)\alpha_c}{g}$$

Corresponding torque on the crankshaft $T_c = F_y \cdot AO$.

or,

$$T_c = \left[\dfrac{W_c l_1 (1-L)\alpha_c}{g}\right] \cdot \left(\dfrac{AO}{AP}\right)$$

Now $AO = OC$ and $AP = CP\cos\phi$

Also,

$$\cos\phi = \left[1 - \sin^2\phi\right]^{0.5} = \dfrac{\left(n^2 - \sin^2\theta\right)^{0.5}}{n}$$

Where,

$$n = \dfrac{CP}{OC} = \dfrac{1}{r}$$

$$T_c = \left[\dfrac{W_c l_1 (1-L)\alpha_c}{g}\right] \cdot \left[\dfrac{\cos\theta}{\left(n^2 - \sin^2\theta\right)^{0.5}}\right]$$

$$= -\left[\frac{W_c l_1 (1-L)}{g}\right] \cdot \left[\frac{\omega^2 (n^2 -1)\sin 2\theta}{2(n^2 -\sin^2 \theta)2}\right]$$

$$= -\left[\frac{W_c l_1 (1-L)}{g}\right] \cdot \left[\frac{\omega^2 \sin 2\theta}{2n^2}\right]$$

Vertical force through $C = W_c \cdot \dfrac{PG}{PC} = \dfrac{W_c l_1}{l}$

Torque exerted on crankshaft by gravity,

$$T_g = -\left(\frac{W_c l_1}{l}\right) \cdot AO$$

$$AO = OC\cos\theta = r\cos\theta$$

$$T_g = -\left(\frac{W_c l_1}{n}\right) \cdot \cos\theta$$

Total torque exertedon the crankshaft by the inertia of moving part,

$$= T_i + T_c + T_e$$

## Graphical Method

The graphical construction for calculating the inertia forces in a reciprocating engine is shown in figure. The following procedure may be adopted for this purpose:

- Draw the acceleration diagram OCQP by Klein's construction. The acceleration of the piston P with respect to the crank Center o is acting in the direction from N to O. Therefore the inertia force $F_i$ shall act in the opposite direction from zero to N.

$$f_p = \omega^2 \cdot NO$$

- Replace the connecting rod by dynamically equivalent system of two masses. Let one of the masses be placed at P. To obtain the position of the other mass, draw GZ perpendicular to CP such that GZ=K, the radius of gyration of the connecting rod. Join PZ and from Z draw perpendicular to DZ which intersects CP at D. Now D is the position of the second mass. Otherwise, GP.GD=k².

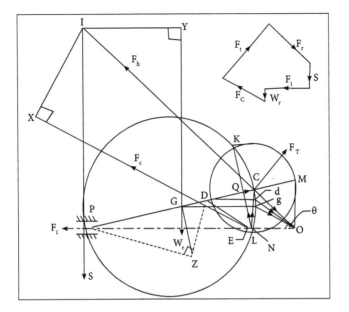

Graphical method to determine inertia forces in reciprocating engine.

- Locate the points g and d on NC, the acceleration image of the rod, by drawing parallel lines from G and D to the line of stroke, Join gO and dO. Then,

$$f_G = \omega^2 \cdot gO \text{ and } f_D = \omega^2 \cdot dO$$

- From D, draw DE parallel to dO to intersect the line of stroke at E. The inertia force of the rod $F_r$ acts through E and in the opposite direction,

$$F_r = m_r \omega^2 \cdot gO$$

Where,

   $m_r$ = The mass of the rod.

The force acting on the connecting rod are:

- Inertia force of the reciprocating parts $F_i$ acting along the line of stroke PO.

- The side thrust between the cross-head and the guide bars S acting at P and right angles to the line of stroke.

- The weight of the connecting rod, W = m.g.

- Inertia force of the connecting rod $F_r$.

- The radial force $F_r$ acting through O and parallel to the crank OC.

- The force F acting perpendicular to the crank OC.

- Now produce the line of action of Fr and S to intersect at a point I. From I draw IX and IY perpendicular to the lines of action of Fr and Wr Taking moments about I, we have,

$$F_t \cdot IC = F_t \cdot IP + F_t \cdot IX + W_r \cdot IY$$

The value of $F_t$ may be obtained from this equation and from the force polygone, the force S and $F_r$ may be calculated. Then, torque exerted on the crankshaft to overcome the inertia of moving parts is = $F_t$.OC.

The following points are to be considered while solving problems by this method:

- Draw the configuration design to a suitable scale.

- Locate all fixed point in a mechanism as a common point in velocity diagram.

- Choose a suitable scale for the vector diagram velocity.

- The velocity vector of each rotating link is rectangular (r) to the link.

- Velocity of each link in mechanism has both magnitude and direction. Start from a point whose magnitude and direction is known.

- The points of the velocity diagram are indicated by small letters.

## Configuration Diagram

It is a skeleton or a line diagram which represents a machine or a mechanism. To study the velocity and acceleration of any mechanisms, first we have to draw the configuration diagram. Configuration diagram is also called as space diagram.

## Problems

1. The following data refer to the lengths of links of a six link mechanism in which the rotary motion of the input link 2 is transformed to the horizontal linear motion of the output slider 6. Fixed link 1, $A_0 B_0$ = 60 mm, Input link 2 $A_0 A$ = 25 mm, Coupler link 3, AB = 85 mm, Follower link 4, $BB_0$ = 55 mm, Connecting rod 5, CD = 60 mm. The pin joint C is at the center of link $BB_0$. The horizontal line of stroke of the slider passes through the fixed link pivots $A_0$ and $B_0$. $B_0 A_0$ A is 60°.In the position:

  i.  Let us sketch the mechanism and indicate the data.

  ii. Let us draw the velocity diagram and determine the linear velocity of the slider, if the input link constant speed is 2 rad/sec. clockwise.

  iii. Let us draw the acceleration diagram and determine the linear acceleration of the slider, which is connected at one end of the connecting rod, CD.

Solution:

Given Data:

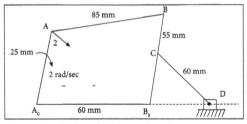

i. Sketch the mechanism.

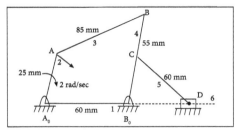

ii. Velocity diagram.

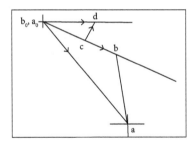

iii. Acceleration diagram.

2. A mechanism of a crank and slotted lever quick return motion is shown in below figure (1) If the crank rotates counter clockwise at 120 r.p.m., let us determine for the configurations shown, the velocity and acceleration of the ram D and also the angular acceleration of the slotted lever. Crank, AB = 150 mm; Slotted arm, OC = 700 mm and link CD = 200 mm.

Solution:

Given:

AB= 150mm, OC = 700mm, CD= 200mm

Rpm = 120

We know that velocity of B with respect to A,

$$V_{BA} = \omega_{BA} \times AB$$

= 12.57 × 0.15 = 1.9 m/s ... (Perpendicular to AB)

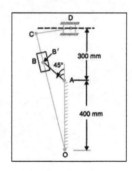

## Acceleration in Mechanisms

By measurement, we find that velocity of the ram D.

$$V_D = \text{Vector at} = 2.15 \text{ m/s}$$

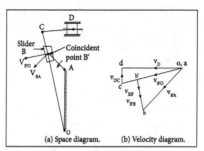

(a) Space diagram (b) Velocity diagram.

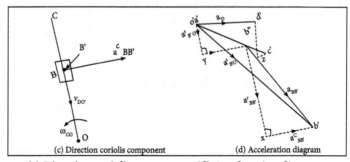

(c) Direction coriolis component (d) Acceleration diagram.

From velocity diagram, we also find that Velocity of B with respect to B,

$$V_{BB}' = \text{Vector b'b} = 1.05 \text{m/s}$$

Velocity of D with respect to C,

$$V_{DC} = \text{Vector cd} = 0.45 \text{ m/s}$$

Velocity of B with respect to O,

$$V_{BO} = \text{vector ob'} = 1.55 \text{ m/s}$$

Velocity of C with respect to O,

$$V_{co} = \text{Vector oc} = 2.15 \text{ m/s}$$

∴ Angular velocity of the link OC or OB,

$$\omega_{co} = \omega_{BO} = \frac{V_{co}}{OC} = \frac{2.15}{0.7} = 3.07 \text{ rad/s (Anti clockwise)}$$

## Acceleration of the Ram D

We know that radial component of the acceleration of B with respect to A,

$$\alpha_{BA}^{\gamma} = \omega_{BA}^2 \times AB = (12.57)^2 \, 0.15 = 23.7 \text{m} / \text{s}^2$$

Coriolis component of the acceleration of slider B with respect to the coincident point B',

$$a^c_{BB}{}' = 2\omega \cdot v = 2\omega_{CO}{}' v_{BB}{}' \quad ...(\because \omega = \omega_{CO} \text{ and } v = v_{BB}{}')$$
$$= 2 \times 3.07 \times 1.05 = 6.45 \text{m} / \text{s}^2$$

Radial component of the acceleration of D with respect to C,

$$a'_{DC} = \frac{v^2_{DC}}{CD} = \frac{(0.45)^2}{0.2} = 1.01 \text{m} / \text{s}^2$$

Radial component of the acceleration of the coincident point B' with respect to O,

$$a^{\gamma}_{B'_o} = \frac{v^2_{BO}}{B'O} = \frac{(.155)^2}{0.52} = 4.62 \text{m} / \text{s}^2 \quad ... \text{ (By measurement B'O = 0.52 m)}$$

By measurement, we find that acceleration of the ram D is,

$$a_D = \text{Vector o' a''} = 8.4 \text{m} / \text{s}^2$$

## Angular Acceleration of the Slotted Lever

By measurement from acceleration diagram, we find that tangential component of the coincident point B' with respect to,

$$a^r_{B'O} = \text{vector y b''} = 6.4 \text{m} / \text{s}^2$$

We know that angular acceleration of the slotted lever,

$$= \frac{a_{B'O}}{OB} = \frac{6.4}{0.52} = 12.3 \text{ rad} / \text{s}^2 \text{ (Anticlockwise)}$$

3. The mechanism of a warping machine is shown in the figure given below. Various dimensions are as follows. $O_1A = 100$ mm; $AC = 700$ mm; $BC = 200$ mm; $BD = 150$ mm; $O_2D = 200$ mm; $O_2E = 400$ mm; $O_3C = 200$ mm. The crank $O_1A$ rotates at a uniform speed of 100 rad/sec. Let us determine the:

   i.   Linear velocity of the point E on the bell crank lever.

   ii.  Angular velocity of links AC and BD.

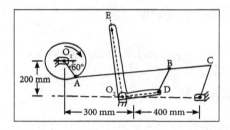

**Solution:**

**Given:**

## Space Diagram: Scale

$1\,cm = 50\,mm$

$\omega_{AO_1} = 100\,rad/s$

$O_1 A = 100\,mm = 0.1\,m = 10\,m/s$

$V_{AO} = \omega_{AO} \times O_1 A = 100 \times 0.1$

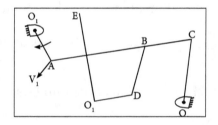

## Velocity Diagram: Scale

$V_{AO1} = 10\,m/s\,1\,cm = 2\,m/s$

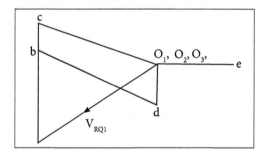

By measurement, we find that the velocity of point C with respect to A.

$V_{CA}$ = Vector ac = 3.5 cm

$= 3.5 \times 2$

$= 7\,m/s$

Velocity of point E on the bell crank lever.

$$V_E - V_{EO2} = \text{Vector } o_2 e = 29.9 \text{ cm}$$

$$= 2.9 \times 2$$

$$= 5.8 \text{ m/s}$$

Velocity of Point D with respect to B,

$$V_{BD} = \text{Vector } bd = 5.1 \text{ cm} = 10.2 \text{ m/s}$$

Angular velocity of link,

$$AC, \omega_{AC} \times \frac{V_{AC}}{A_C} = \frac{7}{0.7}$$

$$= 10 \text{ rad/s}$$

Angular velocity of link,

$$BD, \omega_{BD} = \frac{V_{DB}}{BD} = \frac{10.2}{0.15}$$

$$= 68 \text{ rad/s}$$

4. A four bar chain is represented by a quadrilateral ABCD in which AD is fixed and is 0.6 m long. The crank AB = 0.3 m long rotates in a clockwise direction at 10 rad/s and with an angular acceleration of 30 rad/s' tow both clockwise.

The crank drives the link CD (= 0.36 m) by means of the connecting link BC (= 0.36 m). The angle BAD = 60°. Using graphical method, let us determine the angular velocities and angular accelerations of CD and BC.

Solution:

Given:

Scale: 1 cm = 0.1 m

AD - 0.6 m = 6 cm < BAD - 60°

AB = 0.3 m = 3 cm

CD = 0.36 m = 3.6 cm

BC = 0.36 m = 3.6 cm

$$\omega = 30 \text{ rad}/S^2$$

## Configuration Diagram

Scale 1 cm = 0.1 m

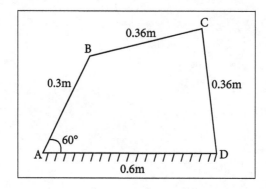

## Angular Velocity

i. $V_{CD} = \omega \times CD$

= 30 × 3.6

$V_{CD} = 36$ m/s

ii. $V_{BC} = 10 \times BC$

= 30 × 3.6

= 36 m/s

## Angular Acceleration

ii. $a_{CD} = \omega^2 \times CD$

= 900 × 3.6

= 3240 m/s²

ii. $a_{BC} = \omega^2 \times BC$

= 900 × 3.6

= 3240 m/s²

## 2.1.2 Piston Effort and Force Acting Along the Connecting Rod

## Piston Effort (F$_p$)

Net force acting on the piston or cross-head pin along the line of stroke is called the piston effort and is denoted by F$_p$ in figure.

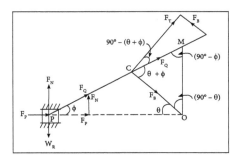

Forces on the reciprocating parts of an engine.

$$F_p = F_L (+/-) F_1 - R_F$$

For a horizontal engine where $F_L$, is the net load acting on the piston. For a single cylinder $p \times A$ and for a double acting $(p_1 A_1 - p_2 A_2)$ or $[p_1 A_1 - p_2 (A_1 - a)]$ where $p, p_1, p_2$ represent pressures and $A, A_1, A_2$ and a represent cross-sectional areas of the cylinder on either side of the piston and a area of cross-section of the piston rod. In case the piston rod 'a' is neglected or negligible, then net load or force on the piston,

$$F_L = (p_1 - p_2) A_1$$

$F_1$ is the inertia force of the reciprocating parts and $R_F$ is the frictional resistance. The inertia forces $F_1$ due to the acceleration of the reciprocating parts opposes the accelerating force $F_p$ on the piston. Use negative sign (-ve) for $0°$ between $0°$ and $180°$, gives acceleration and positive sign (+ve) for $\theta$ between $180°$ and $360°$, gives retardation.

Using the expressions derived for the accelerations above, the inertia force is given by,

$$F_I = M_R \cdot a_R = m_R \omega^2 r \left\{ \cos\theta + \frac{\cos^2 \theta}{n} \right\}$$

Where,

$m_R$ = Mass of the reciprocating parts.

$a_R$ = The acceleration of the reciprocating parts.

$r$ = Crank radius.

$\theta$ = Crank angle.

$$F_p = (F_L \pm F_1 \pm W_R - R_F)$$

for a vertical engine. Here the weight of reciprocating parts assists the piston effort while moving downwards and opposes while moving upwards. So take $+ W_R$ for downward and $-W_R$ for upward.

i. Force along connecting rod $(F_Q)$,

$$F_Q = \frac{F_P}{\cos\phi} = \frac{F_P}{\sqrt{\left[1 - \frac{\sin^2\theta}{n^2}\right]}}$$

From the geometry of the figure and $L\sin\phi = r\sin\theta$ or $\sin\phi = \frac{\sin\theta}{n}; \cos\phi = \sqrt{(1-\sin 2\phi)}$,

$$\cos\phi = \sqrt{1 - \frac{\sin^2\theta}{n^2}}$$

ii. Force along connecting rod $(F_Q)$.

It is denoted by $F_Q$ in above the figure. From the geometry of the figure, we find that,

$$\cos\phi = \sqrt{1 - \frac{\sin^2\theta}{n^2}}$$

$$F_Q = \frac{F_P}{\sqrt{1 - \frac{\sin^2\theta}{n^2}}}$$

From the geometry of the figure and $L\sin\phi$,

$$= r\sin\theta \, or \, \sin\phi = \frac{\sin\theta}{n}; \cos\phi = \sqrt{\left(1 - \sin^2\phi\right)}$$

$$\text{Hence,} \quad \cos\phi = \sqrt{\left[1 - \frac{\sin^2\theta}{n^2}\right]}.$$

## 2.1.3 Crank Effort and Turning Moment on Crankshaft

### Crank Effort

Force is exerted on the crank pin as a result of the force on the piston. Crank effort is the net effort applied at the crank pin perpendicular to the crank which gives the required turning moment on the crankshaft.

Let $F_t$ = crank effort as,

$$F_t \times r = F_c \times r \times \sin(\theta + \beta)$$

$$= \frac{F}{\cos\beta}\sin(\theta + \beta)$$

## Turning Moment on Crankshaft

$$T = F \times r$$

$$\frac{F}{\cos\beta}\sin(\theta+\beta)\times r = \frac{Fr}{\cos\beta}(\sin\theta\cos\beta+\cos\theta\sin\beta)$$

$$= Fr\left(\sin\theta+\cos\theta\sin\beta\frac{I}{\cos\beta}\right)$$

$$= Fr\left(\sin\theta+\cos\theta\frac{\sin\theta}{n}\frac{1}{\frac{1}{n}\sqrt{n^2-\sin^2\theta}}\right)$$

$$= Fr\left(\sin\theta+\frac{2\sin\theta\cos\theta}{2\sqrt{n^2-\sin^2\theta}}\right)$$

$$= Fr\left(\sin\theta+\frac{\sin 2\theta}{2\sqrt{n^2-\sin^2\theta}}\right)$$

Also as r sin (θ+β) = OD cos β,

$$T = F_t \times r$$

$$= \frac{F}{\cos\beta}r\sin(\theta+\beta)$$

$$= \frac{F}{\cos\beta}(OD\cos\beta)$$

$$T = F \times OD$$

## Problems

1. Let us determine the inertia force for the following data of an I.C. Engine. Bore = 175 mm, stroke = 200 mm, engine speed = 500 r.p.m., length of connecting rod =400 mm, crank angle = 60° from T.D.C and mass of reciprocating parts = 180 kg.

Solution:

Given:

D =175 mm

N= 500 r.p.m. Or ω= 2π× 500/60 =52.4 rad/s

L= 200 mm = 0.2 m or r= L/ 2 = 0.1 m

l= 400 mm = 0.4 m; $m_R$= 180 kg

The inertia force may be calculated by graphical method or analytical method as discussed below:

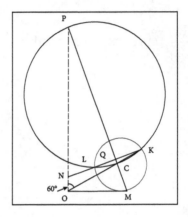

## 1. Graphical Method

First of all, draw the Klien's acceleration diagram OCQN to some suitable scale as shown in figure. By measurement,

ON = 38 mm = 0.038 m

∴ Acceleration of the reciprocating parts,

$$a_R = \omega^2 \times ON = (52.4)2 \times 0.038 = 104.34 \text{ m/s}$$

We know that inertia force,

$$F_i = m_R \times a_R = 180 \times 104.34 \text{ N}$$

$$= 18\,780 \text{ N} = 18.78 \text{ kN}$$

## 2. Analytical Method

We know that ratio of lengths of connecting rod and crank,

n= l/r= 0.4 / 0.1 = 4

$$F_i = m_R \cdot \omega^2 r \left( \cos\theta + \frac{\cos 2\theta}{n} \right)$$

$$= 180 \times (52.4)^2 \times 0.1 \left( \cos 60° + \frac{\cos 120°}{4} \right)$$

$$= 18.52 \text{kN}$$

2. During a trial on steam engine, it is found that the acceleration of the piston is 36 m/s² when the crank has moved 30° from the inner dead centre position. The net effective

steam pressure on the piston is 0.5 N/mm² and the frictional resistance is equivalent to a force of 600 N. The diameter of the piston is 300 mm and the mass of the reciprocating parts is 180 kg. If the length of the crank is 300 mm and the ratio of the connecting rod length to the crank length is 4.5, let us determine:

i. Reaction on the guide bars.

ii. Thrust on the crank shaft bearings.

iii. Turning moment on the crank shaft.

Solution:

Given:

$$a_p = 36 \text{ m/s}^2$$

$$\theta = 30°$$

$$p = 0.5 \text{ N/mm}^2$$

$$R_F = 600 \text{ N, D= 300 mm}$$

$$m_R = 180 \text{ kg}$$

$$r = 300 \text{ mm} = 0.3 \text{ m}$$

$$n = l/r = 4.5$$

## 1. Reaction on the Guide Bars

First of all, let us find the piston effort (F). We know that load on the piston,

$$F_L = p \times \frac{\pi}{4} \times D^2 = 0.5 \times \frac{\pi}{4} \times (300)^2 = 35350 \text{N}$$

and inertia force due to reciprocating parts,

$$F_I = m_R \times a_p = 180 \times 36 = 6480 \text{N}$$

Piston effort: $F_P = F_L = F_I - R_F = 35350 - 6480 - 600 = 28270 \text{N} = 28.27 \text{kN}.$

Let,

$\phi$ = Angle of inclination of the connecting rod to the line of stroke.

We know that $\sin \phi = \sin \theta / n = \sin 30° / 4.5 = 0.1111.$

$$\phi = 6.38°$$

We know that reaction on the guide bars,

$$F_N = F_P \tan \phi = 28.27 \tan 6.38° = 3.16 \text{ kN}$$

## 2. Thrust on the Crank Shaft Bearing

We know that thrust on the crank shaft bearings,

$$F_B = \frac{F_p \cos(\theta + \phi)}{\cos \phi} = \frac{28.27 \cos(30° + 6.38°)}{\cos 6.38°} = 22.9 \text{kN}$$

## 3. Turning moment on the crank shaft

We know that turning moment on the crank shaft,

$$T = \frac{F_p \sin(\theta + \phi)}{\cos \phi} \times r = \frac{28.27 \sin(30° + 6.38°)}{\cos 6.38°} \times 0.3 \text{kN} - \text{m}$$

$$= 5.06 \text{kN-m}$$

3. A vertical, single cylinder, single acting diesel engine has a cylinder diameter 300 mm, stroke length 500 mm, and connecting rod length 4.5 times the crank length. The engine runs at 180 r.p.m. The mass of the reciprocating parts is 280 kg. The compression ratio is 14 and the pressure remains constant during the injection of the oil for 1/10th of the stroke. If the compression and expansion follows the law $p.V^{1.35} = $ Constant, Let us determine,

    i.   Crank-pin effort.

    ii.  Thrust on the bearings.

    iii. Turning moment on the crank shaft, when the crank displacement is 45° from the inner dead centre position during expansion stroke. The suction pressure may be taken as 0.1 N/mm².

Solution:

Given:

        D = 300 mm = 0.3 m

        N= 180 r.p.m.

        L= 500 mm = 0.5 m

        l= 4.5 ro r n= l/r = 4.5

        r= 0.25 m

        $\omega$= 2π× 180/60 = 18.85 rad/s

        $m_R$= 280 kg

        $\dfrac{V_1}{V_2} = 14$

        $\theta = 45°$; $p_1 = 0.1$ N/mm²

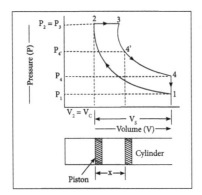

The pressure-volume (i.e. p-V) diagram for a diesel engine is shown in figure in which 1-2 represents the compression, 2-3 represents the injection of fuel, 3-4 represents the expansion, and 4-1 represents the exhaust.

Let,

$p_1$, $p_2$, $p_3$, and $p_4$ = Pressures corresponding to points 1, 2, 3 and 4 respectively,

$V_1$, $V_2$, $V_3$, and $V_4$ = Volumes corresponding to points 1, 2, 3 and 4 respectively.

Since the compression follows the law $p.V^{1.35}$ = constant, therefore,

$$P_1\left(V_1\right)^{1.35} = p_2\left(V_2\right)^{1.35}$$

$$p_2 = p_1\left(\frac{V_1}{V_2}\right)^{1.35} = 0.1\times(14)^{1.35} = 3.526\,\text{N/mm}^2$$

We know that swept volume,

$$V_S = \frac{\pi}{4}\times D^2 \times L = \frac{\pi}{4}\times(0.3)^2 \times 0.5 = 0.035\,\text{m}^3$$

Compression ratio, $= \dfrac{V_1}{V_2} = \dfrac{V_C + V_S}{V_C} = 1 + \dfrac{V_S}{V_C}$    $\dots\left(\because V_2 = V_C\right)$

$$14 = 1 + \frac{0.035}{V_C} \quad \text{or} \quad V_C = \frac{0.035}{14-1} = 0.0027\,\text{m}^3$$

Since the injection of fuel takes place at constant pressure (i.e. $p_2 = p_3$) and continues up to 1/10th of the stroke, therefore volume at the end of the injection of fuel,

$$V_3 = V_C + \frac{1}{10}\times V_S = 0.0027 + \frac{0.035}{10} = 0.0062\,\text{m}^3$$

When the crank displacement is 45° (i.e. when θ = 45°) from the inner dead centre during expansion stroke, the corresponding displacement of the piston (marked by point 4' on the p-V diagram) is given by,

$$x = r\left[(1-\cos\theta)+\frac{\sin^2\theta}{2n}\right] = r\left[(1-\cos 45°)+\frac{\sin^2 45°}{2\times 4.5}\right]$$

$$= 0.25\left[(1-0.707)+\frac{0.5}{9}\right] = 0.087\,m$$

$$V_4' = V_C + \frac{\pi}{4}\times D^2 \times x = 0.0027 + \frac{\pi}{4}\times(0.3)^2 \times 0.087 = 0.0088\,m^2$$

Since the expansion follows the law $p.V^{1.35}$ = constant, therefore,

$$P_3\left(V_3\right)^{1.35} = P_{4'}\left(V_{4'}\right)^{1.35}$$

$$P_{4'} = P_3\left(\frac{V_3}{V_{4'}}\right)^{1.35} = 3.526\left(\frac{0.0062}{0.0088}\right)^{1.35} = 2.2\,N/mm^2$$

Difference of pressures on two sides of the piston,

$$p = p_4 - p_1 = 2.2 - 0.1 = 2.1\,N/mm^2 = 2.1\times 10^6\,N/m^2$$

$\therefore$ Net load on the piston,

$$F_L = p\times\frac{\pi}{4}\times D^2 = 2.1\times 10^6 \times \frac{\pi}{4}\times(0.3)^2 = 148460\,N$$

Inertia force on the reciprocating parts,

$$F_I = m_R \cdot \omega^2 \cdot r\left(\cos\theta + \frac{\cos 2\theta}{n}\right)$$

$$= 280\times(18.85)^2 \times 0.25\left(\cos 45° + \frac{\cos 90°}{4.5}\right) = 17585\,N$$

We know that net force on the piston or piston effort,

$$F_P = F_L - F_I + W_R = F_L - F_I + m_R \cdot g$$

$$148460 - 17\,858 + 280 \times 9.81 = 133622\,N$$

## 1. Crank-Pin Effort

Let,

$\phi$ = Angle of inclination of the connecting rod to the line of stroke.

We know that, $\sin\phi = \sin\theta/n = \sin 45°/4.5 = 0.1571$

$$\therefore \phi = 9.04°$$

We know that crank-pin effort,

$$F_T = \frac{F_P \sin(\theta+\phi)}{\cos\phi} = \frac{133622 \times \sin(45°+9.04°)}{\cos 9.04°} = 109.522N$$

## 2. Thrust on the Bearings

We know that thrust on the bearings,

$$F_B = \frac{F_P.\cos(\theta+\phi)}{\cos\phi} = \frac{133622 \times \sin(45°+9.04°)}{\cos 9.04°}$$

$$= 79.956kN$$

## 3. Turning Moment on the Crankshaft

We know that the turning moment on the crankshaft,

$$T = F_T \times r = 109.522 \times 0.25 = 27.38kN-m$$

# 2.2   Dynamically Equivalent System

In order to determine the motion of a rigid body, under the action of external forces, it is usually convenient to replace the rigid body by two masses placed at a fixed distance apart, in such a way that:

- The sum of their masses is equal to the total mass of the body.

- The centre of gravity of the two masses coincides with that of the body.

- The sum of mass moment of inertia of the masses about their centre of gravity is equal to the mass moment of inertia of the body.

When these three conditions are satisfied, then it is said to be an equivalent dynamical system. Consider a rigid body, having its Centre of gravity at G, as shown in figure.

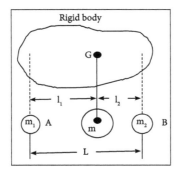

M = Mass of the body,

$k_G$ = Radius of gyration about its Centre of gravity G,

$l_1$ = Distance of mass $m_1$ from G,

$m_1$ and $m_2$ = Two masses which form a dynamical equivalent system,

$l_2$ = Distance of mass $m_2$ from G, and L = Total distance between the masses $m_1$ and $m_2$.

Thus, for the two masses to be dynamically equivalent,

$$M_1 + m_2 = m$$
$$M_1.l_1 = m_2.l_2$$
$$M_1(l_1)^2 + m_2(l_2)^2 = m(k_G)^2$$

From equations (i) and (ii),

$$m_1 = \frac{l_2.m}{l_1 + l_2} \quad \dots i \qquad \dots(i)$$

$$m_2 = \frac{l_1.m}{l_1 + l_2} \quad \dots ii \qquad \dots(ii)$$

Substituting the value of $m_1$ and $m_2$ in equation (iii), we have,

$$\frac{l_2.m}{l_1+l_2}(l_1)^2 + \frac{l_1.m}{l_1+l_2}(l_2)^2 = m(k_G)^2 \text{ or } \frac{l_1 l_2 (l_1\ l_2)}{l_1\ l_2} (k) \quad \dots(iii)$$

$$\therefore \quad l_1 l_2 = (k_G)^2 \qquad \dots(iv)$$

This equation gives the essential condition of placing the two masses, so that the system becomes dynamical equivalent. The distance of one of the masses is arbitrarily chosen and the other distance is obtained from equation (iv).

When the radius of gyration $k_G$ is not known, then the position of the second mass may be obtained by it considering the body as a compound pendulum. The length of the simple pendulum which gives the same frequency as the rigid body is,

$$L = \frac{(k_G)^2 + h^2}{h} = \frac{(k_G)^2 + (l_1)^2}{l_1}$$

We also know that,

$$l_1.l_2 = (k_G)^2$$

$$L = \frac{l_1.l_2 + (l_1)^2}{l_1} = l_2 + l_1$$

This means that the second mass is situated at the centre of oscillation or percussion of the body, which is at a distance of $l_2 = \left(k_G\right)^2 / l_1$.

## Problem

1. A connecting rod is suspended from a point 25 mm above the centre of small end, and 650 mm above its centre of gravity, its mass being 37.5 kg. When permitted to oscillate, the time period is found to be 1.87 seconds. Let us determine the dynamical equivalent system constituted of two masses, one of which is located at the small end centre.

Solution:

Given:

$\quad$ h = 650 mm = 0.65 m

$\quad$ $l_1$ = 650 – 25 = 625 mm = 0.625 m

$\quad$ m = 37.5 kg

$\quad$ $t_p$ = 1.87s

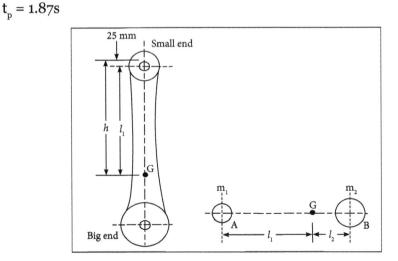

First of all, let us find the radius of gyration ($k_G$) of the connecting rod, about an axis passing through its centre of gravity, G.

We know that for a compound pendulum, time period of oscillation ($t_p$),

$$1.87 = 2\pi\sqrt{\frac{\left(k_G\right)^2 + h^2}{g.h}} \quad \text{or} \quad \frac{1.87}{2\pi} = \sqrt{\frac{\left(k_G\right)^2 + \left(0.65\right)^2}{9.81 \times 0.65}}$$

Squaring both sides, we have,

$$0.0885 = \frac{\left(k_G\right)^2 + 0.4225}{6.38}$$

$$\left(k_G\right)^2 = 0.0885 \times 6.38 - 0.4225 = 0.1425 m^2$$

$$K_G = 0.377 \, m$$

It is given that one of the masses is located at the small end Centre.

Let the other mass is located at a distance $l_2$ from the Centre of gravity G, as shown in figure. We know that, for a dynamically equivalent system,

$$l_1 . l_2 = \left(k_G\right)^2$$

$$l_2 = \frac{\left(k_G\right)^2}{l_1} = \frac{0.1425}{0.625} = 0.228 \, m$$

Let,

m₁ = Mass placed at the small end centre A,

m₂ = Mass placed at a distance $l_2$ from G, i.e. at B.

We know that, for a dynamically equivalent system,

$$m_1 = \frac{l_2 . m}{l_1 + l_2} = \frac{0.228 \times 37.5}{0.625 + 0.228} = 10 \, kg$$

$$m_2 = \frac{l_1 . m}{l_1 + l_2} = \frac{0.625 \times 37.5}{0.625 + 0.228} = 27.5 \, kg$$

### 2.2.1 Compound Pendulum

A simple pendulum is one which can be considered to be a point mass suspended from a string or rod of negligible mass. It is a resonant system with a single resonant frequency. For small amplitudes, the period of such a pendulum can be approximated by:

$$T = 2\pi \sqrt{(L/g)}$$

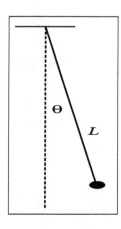

If the rod is not of negligible mass, then it must be treated as a physical pendulum.

The motion of a simple pendulum is like simple harmonic motion in that the equation for the angular displacement is,

$$\theta = \theta_{max} \sin\sqrt{\frac{g}{L}}t$$

Which is the same form as the motion of a mass on a spring:

$$y = A\sin\sqrt{\frac{k}{m}}t$$

The angular frequency of the motion is then given by, $\omega = \sqrt{\frac{g}{L}}$ compared to $\omega = \sqrt{\frac{k}{m}}$ for a mass on spring.

The frequency of pendulum in Hz is given by $f = \frac{1}{2\pi}\sqrt{\frac{g}{L}}$ and the period of motion is then $T = 2\pi\sqrt{\frac{L}{g}}$.

## Centre of Oscillations

If O is the Centre of suspension and G is the centre of gravity of the body then the point O' on OG produced such that OO' = $k^2/h$ (i.e length of the simple pendulum) is known as centre of oscillation of the body.

## Centre of Suspension

The point O, where the perpendicular from the centre of gravity 'G' of the body, cuts the axis of rotation, is called the centre of suspension of the body. Thus the centre of suspension O of a body is the point where the vertical plane through the centre of gravity of the body meets the axis of rotation.

## Compound Pendulum

A compound pendulum also called a physical pendulum, is a body of an arbitrary shape, pivoted at any point so that when the center of mass is displaced on one side, the body starts oscillating in a plane. Unlike a simple pendulum where the entire mass is considered to be situated at the centre of mass, in the case of physical pendulum, we consider the distribution of mass. Let the distance between the pivot and the centre of gravity of the body be l. Then, if the angle of tilt of the pendulum is v, the torque on the body due to the weight of the pendulum acting at the centre of mass is given by,

$$\Gamma = -mgl\sin\varphi$$

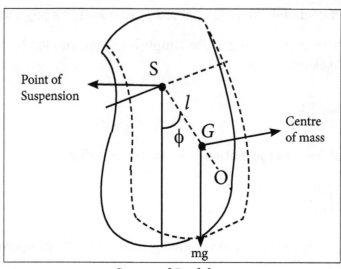

Compound Pendulum

For small $\phi$, $\sin \phi \approx \phi$. The torque acts opposite to the direction of increase of $\phi$. We write the above equation as,

$$\Gamma = -mgl \sin \phi$$
$$= k'\phi$$

Where k' = mgl. Obviously, it is a case of SHM.

Therefore, the time period of oscillation is,

$$T = 2\pi \sqrt{\frac{I}{k'}}$$

Where,

l= Moment of inertia of the compound pendulum around the axis of pivoting and equation. For a physical pendulum becomes,

$$T = 2\pi \sqrt{\frac{l}{mgl}}$$

Defining $L = \dfrac{l}{ml}$

We get,

$$T = 2\pi \sqrt{\frac{L}{g}}$$

Where,

L = Length of an equivalent simple pendulum.

$K$ = Radius of gyration of the compound pendulum through the center of mass, then the moment of inertia, l of the pendulum around an axis passing horizontally through the centre of mass is given by,

$$I_c = mK^2$$

where m is the mass of the pendulum.

Then, according to the theorem of parallel axes, the moment of inertia of the pendulum, l, around the pivot is given by,

$$I = I_c + ml^2$$
$$= m(K^2 + l^2)$$

Hence,

$$T = 2\pi \sqrt{\frac{K^2 + l^2}{gl}}$$

$$= 2\pi \sqrt{\frac{L}{g}}$$

Where,

$$L = \frac{l}{ml} = \frac{K^2 + l^2}{l}$$

Thus if one knows the value of radius of gyration for an irregular body around the axis through the centre of mass, the time period of the oscillation of such a body, can be calculated for different points of pivoting.

## Problem

1. The moment of inertia of a flywheel f mass 90 kg through centre of gravity is 3.6 kg m². This flywheel is suspended in a vertical plane as a compound pendulum. The distance of the centre of gravity from the point of suspension is 120 mm. Let us determine:

   i.   Time period of oscillation.

   ii.  Frequency of oscillation.

Solution:

Given:

   M=90 kg,

   $I_G = 3.6$kg m²

Distance of C.G. From point of suspension,

$$a = 120mm = 0.12m$$

Let,

$$t = \text{Time period of oscillation}$$

$$n = \text{Frequency}$$

We know that $I_G = mk^2$,

$$\therefore \qquad k^2 = \frac{I_G}{m} = \frac{3.6}{90} = 0.04m^2$$

Time period is given by,

$$t = 2\pi\sqrt{\frac{\left(k^2 + a^2\right)}{g \times a}}$$

$$= 2\pi\sqrt{\frac{0.04 + 0.12^2}{9.81 \times 0.12}}$$

$$= 2\pi\sqrt{\frac{0.04 + 0.0144}{9.81 \times 0.12}} \text{ seconds}$$

$$= 1.35 \text{ seconds}$$

Frequency is given by,

$$n = \frac{1}{\ell} = \frac{1}{1.35} = 0.74 \text{ Hertz} = 0.74 \text{ Hz}$$

2. Let us determine the moment inertia of a connecting rod of an engine if weight of the connecting rod is 600 N and its length between the centres is 1 m. The connecting rod oscillates in a vertical plane about the centre of the small end. The distance of its C.G. from the point of oscillation is 700 mm. From the experiment, it is observed that this rod completes 25 oscillation in 48 seconds. And Let us also calculate.

   i.   The moment of inertia foe rod about an axis through the C.C.

   ii.  Length of the equivalent simple pendulum.

Solution:

Given:

$$W = 600N \therefore m = \frac{W}{g} = \frac{600}{9.81} = 61.162 \text{ kg}$$

Length between the centres = 1m.

Distance of C.G. From the point of suspension, a = 700mm = 0.74m.

No. of oscillations = 25 in time 48 sec.

No. of oscillations per second = 25/48 = 0.52.

Frequency, n = 0.52.

For a simple pendulum, the frequency is given by,

$$n = \frac{1}{2\pi}\sqrt{\frac{g}{L}}$$

Length of the simple pendulum which has the same frequency as the given connecting rod is obtained by substituting the value of n=1.92 in the above equation:

$$0.52 = \frac{1}{2\pi}\sqrt{\frac{9.81}{L}}$$

Squaring both sides, we get,

$$(0.52)^2 = \frac{1}{4\pi^2} \times \frac{9.81}{L}$$

$$L = \frac{9.81}{4\pi^2 \times (0.52)^2}$$

$$= 0.9189\text{m}$$

or,

To find $I_G$ first calculate the value of $k^2$. Hence using equation, we get,

$$L = \frac{k^2 + a^2}{a}$$

$$0.9189 = \frac{k^2 + (0.7)^2}{0.7}$$

$$0.9189 \times 1.7 = k^2 + 0.49$$

$$K^2 = 0.9189 \times 0.7 - 0.49 = 0.6432 - 0.49 = 0.15323$$

$$I_G = mk^2$$

$$= 61.162 \times 0.15323 = 9.37 \text{ kg m}^2$$

## 2.2.2 Correction Couple

## Correction Couple Applied to Two Mass System

The location of both of the masses is fixed as it is suitable in case of the relating rod of IC engines. One mass is supposed at the small ending and other at the big ending. In that case all of the equations should not be satisfied and the two mass systems will not be dynamically equivalent to the rigid body. The first two conditions only can be satisfied in this case but third condition will not be satisfied. This condition shall be satisfied if a couple is applied to the two mass systems. This couple is known as the correction couple.

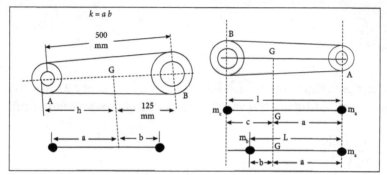

Correction Couple

Shown in figure the connecting rod of an IC engine. This is of length 'l'. It is replaced through two lumped masses $m_a$ and $m_c$ respectively at the piston end and crank pin. In correspondent dynamical system of the connecting rod mass $m_a$ is at the piston end and mass $m_c$ is among the crank pin and centre of gravity 'G'.

As now the location of masses is fixed, hence, the new approximate radius of gyration is given by following,

$$k_1 = a\,c$$

The difference in mass moment of inertia is specified by following,

$$I = I_1 = m(k^2 - k_1^2)$$

If 'α' is angular acceleration of the connecting rod, the difference in inertia torque is given by following,

$$T' = (T - T_1) = m(k_1^2 - k^2)\alpha$$
$$= m\,(ac - ab)\,\alpha = ma\,(c - b)\,\alpha$$

Or,

$$T' = m\,a\,\alpha\,(1 - L)$$

It is the correction couple that should be applied in case, the location of the two masses is fixed. Usually the magnitude of the correction couple is small and may be neglected, whenever, it is needless.

## 2.3 Turning Moment Diagrams for different Types of Engines

Turning moment diagram is a diagram that shows the variation of turning moment or torque acting on crank shaft for various positions of crank. Hence, it is a plot between T and $\theta$. Turning moment is taken along y axis and crank angle $\theta$ is taken along x axis. The inertia effect of connecting rod is generally neglected while drawing these diagrams, but can be considered if required.

Varying torque (T) during one revolution crank-shaft of a steam engine or IC engine is given by,

$$T = F_t \times r$$

$$T = F_r \left[ \sin\theta + \frac{\sin 2\theta}{2\sqrt{n^2 - \sin^2\theta}} \right]$$

Where,

F = Net piston effort

$F_t$ = Crank effort

Crank effort diagram ($F_t$ $V_s$ $\theta$) is identical to a turning moment diagram.

The turning moment diagram for different type of engines are described below.

### Single Cylinder Double Acting Steam Engine

The variation of turning moment with crank angle i for single-cylinder double-acting steam engine is shown in figure.

Turning moment is zero when $\theta = 0°$ and $180°$ and it is maximum when $\theta$ is little less than $90°$.

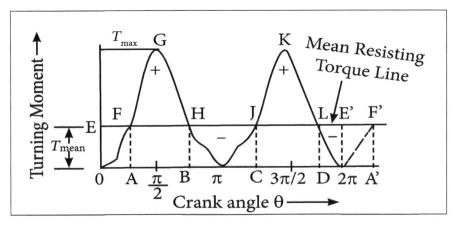

Work done = T. θ

Work done per revolution = T × $2\pi$

= Area of turning moment diagram for one revolution of crank.

That means the area of the turning moment diagram is proportional to the work done per revolution. In the actual practice, the engine is assumed to work against the uniform mean resisting torque.

Mean resisting torque, $T_{mean} = \dfrac{Area\,OGIKQ}{2\pi} = OE$

Area of rectangle,

OEE'Q = OE × OQ

$= T_{mean} \times 2\pi$

= Area OGIKQ

=Work done per revolution

Hence, area of rectangle OEE'Q represents the work done per revolution against mean resisting torque.

Let T = Torque at any instant on the crank shaft.

If T > $T_{mean}$, the engine accelerates and the engine work is stored in the flywheel.

If T < $T_{mean}$ engine retards and flywheel gives up some of its energy to make up the required work.

## Single Cylinder Four-Stroke Internal Combustion Engine

In four-stroke internal combustion engine, the turning moment diagram repeats itself after every two revolutions.

## The four strokes of IC engine are

- Suction stroke for which angle B varies from 0° to 180°.

- Compression stroke for which θ varies from 180° to 380°.

- Expansion stroke for which θ varies from 360° to 540°.

- Exhaust stroke for which θ varies from 540° to 720°.

Figure shows the turning moment diagram for a four stroke internal combustion engine.

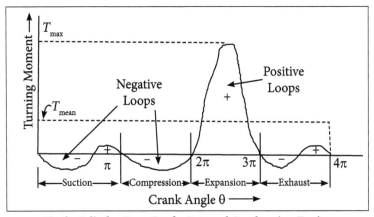

Single Cylinder Four-Stroke Internal Combustion Engine

During suction stroke, the pressure inside the cylinder is less than the atmospheric pressure and hence, the turning moment on the crank is negative for most of the suction stroke. Hence, a negative loop is formed as shown in figure. During compression stroke, the work is done by piston on the gases and hence, a large negative loop is obtained. During expansion stroke the work is done by the gases on the piston and due to this, the turning moment on the crank is positive.

Hence, during this stroke a large positive loop is obtained. Dining exhaust stroke, the work is done by the piston on gases and hence, turning moment on crank is negative for most of the exhaust stroke.

## Multi-Cylinder Engines

For multi-cylinder engines, the total turning moment at any instant is obtained by adding the turning moment developed by each cylinder at that instant. The variation in turning moment is less with more no. of cylinders taken in engine. The speed of the engine will be maximum at crank positions 3, D and F and minimum corresponding to C, E and O.

## Co-efficient of fluctuation of Energy

The ratio between the maximum fluctuations of energy to the work done per cycle is known as the coefficient of fluctuation of energy.

## Problems

1. The turning moment diagram for a multi cylinder engine has been drawn to a scale of 1 mm = 325 Nm vertically and 1 mm = 3° horizontally. The areas above and below the mean torque line are -26,+378, -256, +306, -302, +244, -380, +261 and -225mm. The engine is running at a mean speed of 800 rpm. The total fluctuation of speed is not to exceed ± 1.6% of the mean speed. If the radius of flywheel is 0.7 m, with these conditions we will calculate the mass of the flywheel.

Solution:

Given:

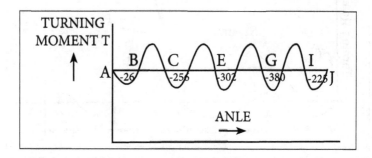

Horizontal scale: 1 mm = 3°

Vertical scale: 1 mm = 325 Nm

Max, fluctuation of speed: $C_s = \pm 1.6\%$

$$= 0.032$$

Mean speed (N) = 800 rpm

Radius of gyration k= 0.7m

Energy at A = $E_A$

Energy at B = $E_A$ −26

Energy at C = $E_A$ + 352

Energy at D = $E_A$ + 96

Energy at E = $E_A$ + 402

Energy at F = $E_A$ + 100

Energy at G = $E_A$ + 344

Energy at H = $E_A$ − 36

Energy at I = $E_A$ + 225

Energy at J = 0

Maximum Energy = $E_A$ + 402

Minimum Energy = $E_A$ − 36

∴ Maximum fluctuation of energy = $(E_A + 402) - (E_a - 36)$ = 438 units.

Converting to absolute units,

$$1\,mm^2 = 3 \times \frac{\pi}{180} \times 325\,N-m$$

$$\therefore \Delta E = 438 \times \frac{\pi}{60} \times 325 = 7{,}453\,N-m$$

Using the formula,

$$I = mk^2 = \frac{\Delta E}{\omega^2 C_S} = \frac{7{,}453}{83.78^2 \times 0.032}$$

$$= 33.18\,kg-m^2$$

$$\left( \omega = \frac{2\pi N}{60} = \frac{2\pi \times 800.}{60} = 83.78\,rad/sec \right)$$

$$C_s = 0.032$$

$$k = \sqrt{\frac{1}{m}}$$

For fly wheel, I,

$$= \frac{mr^2}{2}$$

$$= \frac{m \times 0.7^2}{2}$$

$$\therefore k = \frac{0.7}{\sqrt{2}} = 0.495m$$

$$\therefore \text{Required mass of fly wheel} = \frac{I}{k^2} = \frac{33.18}{0.495^2}$$

$$= 135.4\,kg$$

2. The torque delivered by two-stroke engine is represented by T = (1000 + 300 sin 2θ − 500 cos 2θ) Nm Where θ is the angle turned by the crank from the inner-dead center. The engine speed is 250 rpm. The mass of the flywheel is 400 kg and radius of gyration 400 mm. Here we will determine the power developed, the total percentage fluctuation of speed; the angular acceleration of flywheel when the crank has rotated through an angle of 60° from the inner-dead center and the maximum angular acceleration and retardation of the flywheel.

Solutions:

Given:

$$T = (1000 + 300 \sin 2\theta - 500 \cos 2\theta)\,N-m$$

Where,

$\theta$ – Angle turned by the crank from IDC.

N = 250 rpm.

Mass of flywheel = 400 kg.

Radius of gyration = 400 mm.

Mean torque,

$$T_{mean} = 1/\pi \int_0^\pi T \cdot d\theta$$

$$= \frac{1}{\pi} \int_0^\pi (1000 + 300 \sin 2\theta - 500 \cos 2\theta) d\theta$$

= 1000 N-m

Power = T × W

= 1000 × (2π × 250) 60

= 26180 W

Power = 26.18 kW

Total Fluctuation of Speed,

$\Delta T = T - T_{mean}$

= (1000 + 300 sin 2θ – 500 cos 2θ) – 1000

when ΔT = 0,

300 sin 2θ – 500 cos 2θ = 0

tan 2θ = 5/3

θ = 29.5°

or,

θ = 119.5°

$$e_{max} = \frac{1}{\pi} \int_{29.5°}^{119.5°} \Delta T \, d\theta = 1/\pi \int_{29.5}^{119.5} (300 \sin 2\theta - 500 \cos 2\theta) \, d\theta$$

= 5.83 N –m.

$$k = \frac{e_{max}}{mk^2 \omega^2}$$

$$= \frac{583.1}{400 \times (0.4)^2 \times (2\pi \times 250/60)^2}$$

$$k = 1.329\%$$

Acceleration is produced by excess $\Delta T$ value.

So,

$$\Delta T = 300 \sin 2\theta - 500 \cos 2\theta; \text{ When } \theta = 60°$$

$$= 300 \sin 2(60) - 500 \cos 2(60)$$

$$\Delta T = 509.8 \text{ N-m}$$

We know that,

$$\Delta T = mk^2\alpha$$

$$509.8 = 400 \times (0.4)^2\alpha$$

$$\alpha = 7.95 \text{ rad/s2}$$

for, $\Delta T_{max} = \Delta T_{min}$

$$\frac{d}{d\theta}(\Delta T) = \frac{d}{d\theta}(300 \sin 2\theta - 500 \cos 2\theta)d\theta$$

$$= 2\theta = 149.04° \text{ and } 329.04°$$

When,

$$2\theta = 329°, T = 416 \text{ N-m}, \Delta T = -583.1 \text{ N-m}$$

$$\Delta T = mk^2\alpha$$

$$583 = 400 \times (0.4)^2 \times \alpha$$

$$\alpha = 9.11 \text{ rad/s}^2$$

- Power developed (e) = 26 kw.
- Fluctuation of speed = 1.329%.
- Angular Acceleration = 7.9 rad/s².
- Maximum Angular Acceleration = 9.11 rad/s².

## Fly Wheels and their Design

A flywheel is a rotating mechanical device that is used to store rotational energy. Flywheels have a significant moment of inertia and thus resist changes in rotational speed.

The amount of energy stored in a flywheel is proportional to the square of its rotational speed. Energy is transferred to a flywheel by applying torque to it, thereby increasing its rotational speed and hence it stores energy.

Conversely, a flywheel releases stored energy by applying torque to a mechanical load, thereby decreasing the flywheel's rotational speed.

Single cylinder engine needs a large size flywheel:

Large size flywheel is required in single cylinder since maximum fluctuation of energy is more compared to multi cylinder engines, as evident from turning moment diagrams.

3. The vertical double acting steam engine develops 75kW at 250 rpm. The maximum fluctuation of energy is 30 percent of the work done per stroke. The maximum and minimum speeds do not to vary more than 1 percent on either side of the mean speed. Let us calculate the mass of the flywheel required, if the radius of gyration is 0.6m.

Solution:

Given:

$$\text{Power} = 75\text{kW}$$

$$\text{Rpm} = 250$$

$$\Delta E = 0.3 \times \text{W.D/Stroke}$$

$$C_s = 2\%$$

$$k = 0.6\text{m}$$

Formula used:

$$\text{Work done/Cycle} = \frac{p \times 60}{n}$$

n = Number of working strokes/min.

P = Power developed in Watts.

$$\therefore \text{W.D/Cycle} = \frac{\left(75 \times 10^3\right) \times 60}{250}$$

$$= 18,000 \text{ N-m}$$

$$\text{W.Q/Cycle TB}_{\text{Moon}} \times 2\pi$$

$$\therefore T_{Moon}$$

$$= \frac{18,000}{2}$$

$$= 2700 \text{ N} - \text{m}$$

Maximum fluctuation of energy = $\Delta E$ = 0.3 × W.D/Stroke.

$$0.3 \times \frac{18,000}{2} = 2700 \text{ N} - \text{m}$$

Using the formula,

$$\Delta E = I_\omega^2 C_s,$$

$$I = \frac{\Delta E}{\omega^2 C_S}$$

$$\omega = \frac{2\pi N}{60} = \frac{2\pi \times 250}{60} = 26.18 \text{ rad / sec}$$

$$\therefore mk^2 = \frac{2700}{26.18^2 \times 0.02}$$

$$\therefore m = \frac{2700}{0.6^2 \times 26.18^2 \times 0.02}$$

m= 547 kg

## 2.3.1 Fluctuation of eEnergy and Fluctuation of Speed

## Coefficient of Fluctuation of Energy

"The ratio of the maximum fluctuation of energy to the work done per cycle is defined as the coefficient of fluctuation of energy."

- Let, W = Work done per cycle (Nm) and $K_e$ = Coefficient of fluctuation of energy.

Then, $k_e = \frac{e_{max}}{W}$

Where work done per cycle is the area under T- o curve in the turning moment diagram.

The symbol used for the coefficient of fluctuation of energy is $\beta$,

$$\therefore \beta = \frac{\frac{1}{2}I\omega_2^2 - \frac{1}{2}I\omega_1^2}{W}$$

$$\beta = \frac{I\left(\omega_2^2 - I\omega_1^2\right)}{2W}$$

Since it is a ratio of like quantities, the coefficient of fluctuation of energy $\beta$ has no unit,

Now $\omega_2^2 - \omega_1^2 = (\omega_2 + \omega_1)(\omega^2 - \omega_1)$.

Provide $\omega_2^2 - \omega_1^2 = (\omega_2 + \omega_1)(\omega_2 - \omega_1)$ is small then, from equation $(\omega_2 + \omega_1) = 2\omega$ and, from equation $(\omega_2 - \omega_1) = \phi\omega$;

$$(\omega_2^2 - \omega_1^2) = 2\phi\omega^2$$

$$\beta = \frac{I \times 2\phi\omega^2}{2W}$$

$$I = \frac{\beta W}{\phi\omega^2}$$

## Coefficient of Fluctuation of Speed

"The difference between the greatest speed and the least speed is known as maximum fluctuation of speed and the ratio of the maximum fluctuation of speed to the mean speed is the coefficient of fluctuation of speed."

Let,

$\omega_1$ = Maximum speed of flywheel.

$\omega_2$ = Minimum speed of flywheel.

$\omega$ = Mean speed of flywheel.

q = Coefficient of fluctuation of speed.

$K_s$ = Co-efficient of fluctuation speed, expressed in percentages.

$$\omega_1 = \frac{\omega_1 + \omega_2}{2}$$

$$q = \frac{\omega_1 - \omega_2}{\omega}$$

$$k_s = \frac{\omega_1 - \omega_2}{\omega} \times 100$$

## Problem

A flywheel is required to rotate at a mean speed of 300 rev/min with a fluctuation of speed of 2 % of the mean speed. If the work done per cycle is 40 k J and the coefficient of fluctuation of energy is 0.05, let us determine the mass of the flywheel required if its radius of gyration is 450 mm.

Solution:

Given:

$$I = mk^2 = \frac{\beta W}{\phi \omega^2}$$

$$\therefore m = \frac{\beta W}{\phi \omega^2 k^2}$$

Where,

$\beta = 0.05$

$w = 40kj = 40 \times 10^3$ J

$\phi = \pm 2\% = 0.04$

K=450mm=0.45m

and $\omega = \dfrac{300\, rev/min \times 2\pi\, rad/rev}{60s/min} = 31.42\, rad/s$

$$\therefore m = \frac{0.05 \times 40 \times 10^3\, J}{0.04 \times (31.42\, rad/s)^2 \times (0.45m)^2}$$

=250 kg

i.e. The mass of the flywheel is 250 kg.

## 2.4   Friction of a Screw: Nut and Square Threaded Screw

### External Threads

The screws, bolts, studs, nuts etc. are widely used in various machines and structures for temporary fastenings. These fastenings have screw threads, which are made by cutting a continuous helical groove on a cylindrical surface. The threads are cut on the outer surface of a solid rod.

### Internal Threads

The threads are cut on the internal surface of a hollow rod. The screw threads are mainly of two types:

- V-threads

- Square threads

The V-threads are stronger and offer more frictional resistance to motion than square threads. Moreover, the V-threads have an advantage of preventing the nut from slackening. In general, the V- threads are used for the purpose of tightening pieces together e.g. bolts and nuts etc. But the square threads are used in screw jacks, vice screws etc. The following terms are important for the study of screw.

1. Helix: It is the curve traced by a particle, while describing a circular path at a uniform speed and advancing in the axial direction at a uniform rate. In other words, it is the curve traced by a particle while moving along a screw thread.

2. Pitch: It is the distance from a point of a screw to a corresponding point on the next thread, measured parallel to the axis of the screw.

3. Lead: It is the distance, a screw thread advances axially in one turn.

4. Depth of thread: It is the distance between the top and bottom surfaces of a thread (also known as crest and root of a thread).

5. Single-threaded screw: If the lead of a screw is equal to its pitch, it is known as single threaded screw.

6. Multi-threaded screw: If more than one thread is cut in one lead distance of a screw, it is known as multi-threaded screw e.g. in a double threaded screw, two threads are cut in one lead length. In such cases, all the threads run independently along the length of the rod. Mathematically,

Lead = Pitch × Number of threads

7. Helix angle: It is the slope or inclination of the thread with the horizontal. Mathematically,

$= p/d$...(In single-threaded screw)

$= n.p/d$...(In multi-threaded screw)

Where,

p = Pitch of the screw,

d = Mean diameter of the screw,

n = Number of threads in one lead.

## Self Locking Screw

A screw will be self locking if the friction angle is greater than helix angle or coefficient of friction is greater than tangent of helix angle.

## Square Thread

The square threads, because of their high efficiency, are widely used for transmission of power in either direction. Such types of threads are usually found on the feed mechanisms of machine tools, valves, spindles, screw jacks etc. The square threads are not as strong as V-threads but they offer less frictional resistance to motion than Whitworth threads.

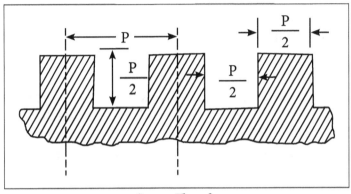

Square Threads

The pitch of the square thread is often taken twice that of a B.S.W. thread of the same diameter. The proportions of the thread are shown in Figure.

### 2.4.1 V-Threaded Screw

We have seen Art that the normal reaction in case of a square threaded screw is,

$R_N$ = W cos α, where α = Helix angle.

But in case of V-thread (or acme or trapezoidal threads), the normal reaction between the screw and nut is increased because the axial component of this normal reaction must be equal to the axial load W, as shown in Figure.

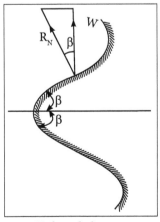

V-threaded screw

Let $2\beta$ = Angle of the V-thread, and $\beta$ = Semi-angle of the V-thread,

$$R_N = \frac{W}{\cos \beta}$$

Frictional force, $F = \mu R_N = \mu \times \dfrac{W}{\cos \beta} = \mu_1 W$

$$\frac{\mu}{\cos \beta} = \mu_1$$

known as virtual coefficient of friction.

## Problem

1. Two co-axial rods are connected by a turn buckle which consists of a box nut, the one screw being right handed and the other left handed on a pitch diameter of 22 mm, the pitch of thread being 3 mm. The included angle of the thread is 60°. Assuming that the rods do not turn, let us calculate the torque required on the nut to produce a pull of 40 kN, given that the coefficient of friction is 0.15.

Solution:

Given:

$d = 22$ mm; $p = 3$ mm

$2\beta = 60°$ or $\beta = 30°$

$W = 40$ kN $= 40 \times 10^3$ N

$\mu = 0.15$

We know that,

$$\tan \alpha = \frac{p}{\pi d} = \frac{3}{\pi \times 22} = 0.0434$$

and virtual coefficient of friction,

$$\mu_1 = \tan \phi_1 = \frac{\mu}{\cos \beta} = \frac{0.15}{\cos 30°} = 0.173$$

We know that the force required at the circumference of the screw,

$$P = W \tan(\alpha + \phi_1) = W \left[ \frac{\tan \alpha + \tan \phi_1}{1 - \tan \alpha . \tan \phi_1} \right]$$

$$= 40 \times 10^3 \left[ \frac{0.0434 + 0.173}{1 - 0.0434 \times 0.173} \right] = 8720 \, N$$

and torque on one rod, $T = P \times d/2 = 8720 \times 22/2 = 95\,920$ N-mm $= 95.92$ N-m.

Since the turn buckle has right and left hand threads and the torque on each rod is,

T = 95.92 N-m

therefore the torque required on the nut,

$T_1$ = 2T = 2 × 95.92 = 191.84 N-m

2. The mean diameter of a Whitworth bolt having V-threads is 25 mm. The pitch of the thread is 5 mm and the angle of V is 55°. The bolt is tightened by screwing a nut whose mean radius of the bearing surface is 25 mm. If the coefficient of friction for nut and bolt is 0.1 and for nut and bearing surfaces 0.16. Let us determine the force required at the end of a spanner 0.5 m long when the load on the bolt is 10 kN.

Solution:

Given:

d = 25 mm; p = 5 mm; 2 β = 55° or β = 27.5°;

R = 25 mm; μ = tan φ = 0.1; $μ_2$ = 0.16;

l = 0.5 m; W = 10 kN = 10 × 10³ N

We know that virtual coefficient of friction,

$$\mu_1 = \tan\phi_1 = \frac{\mu}{\cos\beta} = \frac{0.1}{\cos 27.5°} \frac{0.1}{0.887} = 0.113$$

$$\tan\alpha = \frac{p}{\pi d} = \frac{5}{\pi \times 25} = 0.064$$

∴ Force on the screw,

$$P = W\tan(\alpha + \phi_1) = W\left[\frac{\tan\alpha + \tan\phi_1}{1 - \tan\alpha.\tan\phi_1}\right]$$

$$= 10 \times 10^3\left[\frac{0.064 + 0.113}{1 - 0.064 \times 0.113}\right] = 1783\,N$$

We know that total torque transmitted,

$$T = P \times \frac{d}{2} + \mu_2 W.R = 1783 \times \frac{25}{2} + 0.16 \times 10 \times 10^3 \times 25\ N-mm$$

= 62300 N-mm 62.3 N-m                              ...(i)

Let $P_1$ = Force required at the end of a spanner.

$\therefore$ Torque required at the end of a spanner,

$$T = P_1 \times 1 = P_1 \times 0.5 = 0.5\, P_1 \text{ N-m} \qquad\qquad ...(ii)$$

Equating equations (i) and (ii),

$$P_1 = 62.3/0.5 = 124.6 \text{ N}$$

## 2.4.2 Pivot and Collar Friction

The rotating shafts are frequently subjected to axial thrust. The bearing surfaces such as pivot and collar bearings are used to take this axial thrust of the rotating shaft. The propeller shafts of ships, the shafts of steam turbines and vertical machine shafts are examples of shafts which carry an axial thrust.

The bearing surfaces placed at the end of a shaft to take the axial thrust are known as pivots. The pivot may have a flat surface or conical surface as shown in figure (a) and (b) respectively. When the cone is truncated, it is then known as truncated or trapezoidal pivot as shown in figure (c).

The collar may have flat bearing surface or conical bearing surface, but the flat surface is most commonly used. There may be a single collar, as shown in figure (d) or several collars along the length of a shaft, as shown in figure (e) in order to reduce the intensity of pressure.

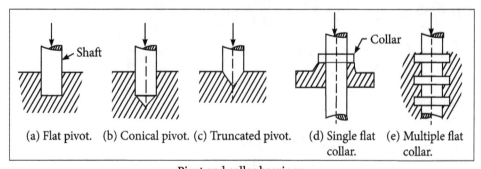

(a) Flat pivot.   (b) Conical pivot. (c) Truncated pivot.   (d) Single flat   (e) Multiple flat
                                                                  collar.           collar.

Pivot and collar bearings.

In modern practice, ball and roller thrust bearings are used when power is being transmitted and when thrusts are large as in case of propeller shafts of ships.

A little consideration will show that in a new bearing, the contact between the shaft and bearing may be good over the whole surface. In other words, we can say that the pressure over the rubbing surfaces is uniformly distributed. But when the bearing becomes old, all parts of the rubbing surface will not move with the same velocity, because the velocity of rubbing surface increases with the distance from the axis of the bearing.

This means that wear may be different at different radii and this will alter the distribution of pressure. Hence, in the study of friction of bearings, it is assumed that:

- The pressure is uniformly distributed throughout the bearing surface.
- The wear is uniform throughout the bearing surface.

## Flat Pivot Bearing

When a vertical shaft rotates in a flat pivot bearing as shown in figure, the sliding friction will be along the surface of contact between the shaft and the bearing.

Let,

$W$ = Load transmitted over the bearing surface,

$R$ = Radius of bearing surface,

$p$ = Intensity of pressure per unit area of bearing surface between rubbing surfaces,

$\mu$ = Coefficient of friction.

We will consider the following two cases:

- When there is a uniform pressure.
- When there is a uniform wear.

## 1. Considering Uniform Pressure

When the pressure is uniformly distributed over the bearing area, then:

$$p = \frac{W}{\pi R^2}$$

Consider a ring of radius r and thickness dr of the bearing area.

$\therefore$ Area of bearing surface, $A = 2\pi r.dr$,

Load transmitted to the ring,

$$\delta W = p \times A = p \times 2\pi r.d$$

Frictional resistance to sliding on the ring acting tangentially at radius r,

$$F_r = \mu.\delta W = \mu p \times 2\pi r.dr = 2\pi \mu.p.r.dr$$

$\therefore$ Frictional torque on the ring,

$$T_r = F_r \times r = 2\pi \mu p r.dr \times r = 2\pi \mu p r^2 dr$$

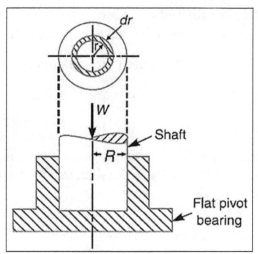

Flat pivot or Footstep bearing.

Integrating this equation within the limits from 0 to R for the total frictional torque on the pivot bearing.

∴ Total frictional torque,

$$T = \int_0^R 2\pi\mu\, p r^2 dr = 2\pi\mu\, p \int_0^R r^2 dr$$

$$= 2\pi\mu\, p \left[\frac{r^3}{3}\right]_0^R = 2\pi\mu\, p \times \frac{R^3}{3} = \frac{2}{3} \times \pi\mu.p.R^3$$

$$= \frac{2}{3} \times \pi\mu \times \frac{W}{\pi R^2} \times R^3 = \frac{2}{3} \times \mu.W.R$$

When the shaft rotates at $\omega$ rad/s, then power lost in friction,

$$P = T.\omega = T \times 2\pi\, N/60 \left(\omega = 2\pi\, N/60\right)$$

where N = Speed of shaft in r.p.m.

## 2. Considering Uniform Wear

We have already discussed that the rate of wear depends upon the intensity of pressure (p) and the velocity of rubbing surfaces (v). It is assumed that the rate of wear is proportional to the product of intensity of pressure and the velocity of rubbing surfaces (i.e. p.v.). Since the velocity of rubbing surfaces increases with the distance (i.e. radius r) from the axis of the bearing, therefore for uniform wear.

p.r = C (a constant) or p = C / r

and the load transmitted to the ring,

$$\delta W = p \times 2\pi r.dr$$

$$= \frac{C}{r} \times 2\pi r.dr = 2\pi C.dr$$

∴ Total load transmitted to the bearing,

$$W = \int_0^R 2\pi C.dr = 2\pi C[r]_0^R = 2\pi C.R \ \text{ or } \ C = \frac{W}{2\pi R}$$

We know that frictional torque acting on the ring,

$$T_r = 2\pi\mu\, pr^2 \ dr = 2\pi\mu \times \frac{C}{r} \times r^2 dr$$

$$= 2\pi\mu.C.r\, dr$$

∴ Total frictional torque on the bearing,

$$T = \int_0^R 2\pi\mu.C.r.dr = 2\pi\mu.C\left[\frac{r^2}{2}\right]_0^R$$

$$= 2\pi\mu.C \times \frac{R^2}{2} = \pi\mu.C.R^2$$

$$= \pi\mu \times \frac{W}{2\pi R} \times R^2 = \frac{1}{2} \times \mu.W.R \quad \cdots\left(\because C = \frac{W}{2\pi R}\right)$$

## Flat Collar Bearing

The collar bearings are used to take the axial thrust of the rotating shafts. There may be a single collar or multiple collar bearings as shown in figure (a) and (b) respectively. The collar bearings are also called as thrust bearings. The friction in the collar bearings may be found as explained below:

Consider a single flat collar bearing supporting a shaft as shown in figure (a).

Let,

$r_1$ = External radius of the collar.

$r_2$ = Internal radius of the collar.

∴ Area of the bearing surface.

$$A = \pi\left[\left(r_1\right)^2 - \left(r_2\right)^2\right]$$

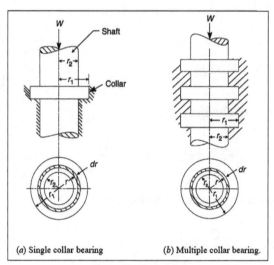

(a) Single collar bearing      (b) Multiple collar bearing.

Flat collar bearings

## 1. Considering Uniform Pressure

When the pressure is uniformly distributed over the bearing surface, then the intensity of pressure,

$$p = \frac{W}{A} = \frac{W}{\pi\left[\left(r_1\right)^2 - \left(r_2\right)^2\right]}$$

The frictional torque on the ring of radius r and thickness dr is,

$$T_r = 2\pi\mu.p.r^2.dr$$

Integrating this equation within the limits from r2 to r1 for the total frictional torque on the collar.

∴ Total frictional torque,

$$T = \int_{r_2}^{r_1} 2\pi\mu.p.r^2.dr = 2\pi\mu.p\left[\frac{r^3}{3}\right]_{r_2}^{r_1} = 2\pi\mu.p\left[\frac{\left(r_1\right)^3 - \left(r_2\right)^3}{3}\right]$$

Substituting the value of p from equation above,

$$T = 2\pi\mu \times \frac{W}{\pi\left[\left(r_1\right)^2 - \left(r_2\right)^2\right]}\left[\frac{\left(r_1\right)^3 - \left(r_2\right)^3}{3}\right]$$

$$= \frac{2}{3}\times\mu.W\left[\frac{\left(r_1\right)^3 - \left(r_2\right)^3}{\left(r_1\right)^2 - \left(r_2\right)^2}\right]$$

## 2. Considering Uniform Wear

The load transmitted on the ring, considering uniform wear is,

$$\delta W = p_r.2\pi r.dr = \frac{C}{r} \times 2\pi r.dr = 2\pi C.dr$$

∴ Total load transmitted to the collar,

$$W = \int_{r_2}^{r_1} 2\pi C.dr = 2\pi C\left[r\right]_{r_2}^{r_1} = 2\pi C\left(r_1 - r_2\right)$$

Or, $C = \dfrac{W}{2\pi\left(r_1 - r_2\right)}$

We also know that frictional torque on the ring,

$$Tr = \mu.\delta W.r = \mu \times 2\pi C.dr.r = 2\pi\mu.C.r.dr$$

∴ Total frictional torque on the bearing,

$$T = \int_{r_2}^{r_1} 2\pi\mu C.r.dr = 2\pi\mu.C\left[\frac{r^2}{2}\right]_{r_2}^{r_1} = 2\pi\mu.C\left[\frac{\left(r_1\right)^2 - \left(r_2\right)^2}{2}\right]$$

$$= \pi\mu.C\left[\left(r_1\right)^2 - \left(r_2\right)^2\right]$$

Substituting the value of C from equation above,

$$T = \pi\mu \times \frac{W}{2\pi\left(r_1 - r_2\right)}\left[\left(r_1\right)^2 - \left(r_2\right)^2\right] = \frac{1}{2} \times \mu.W\left(r_1 + r_2\right)$$

### 2.4.3 Friction Circle and Friction Axis

Mechanism is an assemblage of kinematic links which are joined by kinematic pair. The turning pair is used kinematic pair in which a circular pin supported in a bearing allows turning or revolving motion between two links. When the pin is at rest in bearing and contact surfaces are frictionless, the reaction of the bearing on the pin will act at point A in upward direction, as shown in the figure (a) below. Hence, when the pin rotates, say in counter clockwise direction, the point of contact is shifted towards left to point B. At this point, the pin is subjected to two faces besides external forces F:

- Normal reaction $R_n$.

- Frictional force $\mu R_n$, which acts in the direction opposite to that of rotation and is tangential at point B.

- The resultant reaction R is inclined at an angle $\phi$ with normal reaction $R_n$.

Therefore, the pin is subjected to two forces-external force F and reaction force R. For equilibrium, the forces F and R should be equal in magnitude. These two forces are parallel to each other at distance OC, hence they constitute a couple.

Frictional torque: $T = F \times OC$

$$= F \times r \sin \phi.$$

For small angle, $\sin \phi \approx \tan \phi = \mu$,

Therefore, $T = \mu Fr$.

Thus the effect of friction on a turning pair is to displace the reaction force through a distance equal to $r \sin \phi = \mu r$ such that it is tangential to circle drawn with radius OC. This circle is commonly called as friction circle.

## Friction Axis

When a link of mechanism is joined to another link by a frictionless turning pairs, the line of action of the force passes through the line joining the centre of the pin but it is tangential to the friction circle of the pin. Hence the line of action of the force is common tangent to the friction circles of two pins. Since there are four possible common tangents to a given pair of friction circles, it is necessary to determine the information:

- The direction of rotation of one link relative to other connecting link.

- The direction of external force acting on the link.

Based on the information obtained above, a common tangent, which gives a friction couple in the direction opposite to that of a couple producing the motion, is selected. This line of action of the force is known as friction axis of the link.

The direction of motion of the link relative to another link can be determined by following criteria.

If the angle between the links at pin joint increases:

1. When two links are connected by a pin joint and driving link rotates in clockwise direction:

- If the angle between the links at pin joint increases, the driven link rotates in counterclockwise direction.

- If the angle at joint decreases, the driven link rotates in clockwise direction.

2. When driving link rotates in counterclockwise direction and angle between the links increases, the driven link rotates in clockwise direction. If the angle decreases, the driven link rotates in counterclockwise direction.

To illustrate, let us consider a four bar mechanism as shown in the figure (a). When the driving link, AB, rotates in clockwise direction, the coupler link BC is subjected to compressive axial force and the angle between these links ∠ABC at point B is increasing, therefore the coupler link BC will rotate in counterclockwise direction relative to driving link AB.

Therefore tangent to the friction circle at point B will be on the upper side to give clockwise friction couple. Similarly, when link CD is rotating in clockwise direction, the angle ∠DCB decreases and the link BC relative to CD will rotate in clockwise direction. Therefore, tangent to the friction circle will be on the lower side to give counterclockwise friction couple. The tangent to these two circles is known as friction axis.

Referring to the another position of a four bar mechanism as shown in the figure (b), the coupler link BC is subjected to tension force and angle ∠ABC at point B is increasing, so link BC will rotate in counterclockwise direction and the friction couple will be on the lower side of the link.

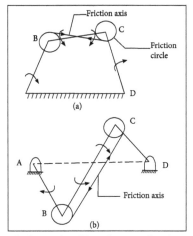

Friction axis of four bar mechanism.

Similarly, with reference to link CD, the coupler BC rotates in clockwise direction such that the friction couple at pin C opposes the motion. The resulting friction axis of the link BC is shown in the figure (b).

The friction axes of slider crank mechanism for various positions of crank are shown in the figure below. In slider crank mechanism, there is another method to know the position of friction axis. Accordingly, out of four possible tangents one that gives the least intercept (OC) from origin O is the required friction axis of the coupler link.

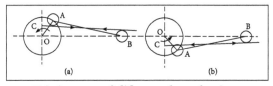

Friction axes of slider crank mechanism.

## Rolling Friction (Circular)

It is the friction experience between surfaces which has balls or rollers interposed between them.

Rolling friction is the resistive force that slows down the motion of a rolling ball or wheel. This type of friction is typically a combination of several friction forces at the point of contact between the wheel and the ground or other surface.

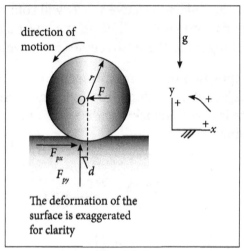

Rolling friction

## Example

Wheels of a locomotive.

### 2.4.4 Friction Clutches

A friction clutch has its principal application in the transmission of power of shafts and machines which must be started and stopped frequently. Its application is also found in cases in which power is to be delivered to machines partially or fully loaded. The force of friction is used to start the driven shaft from rest and gradually brings it up to the proper speed without excessive slipping of the friction surfaces.

In automobiles, friction clutch is used to connect the engine to the driven shaft. In operating clutch, care should be taken so that the friction surfaces engage easily and gradually brings the driven shaft to proper speed. The proper alignment of the bearing should be maintained and it should be located as close to the clutch as possible. It may be noted that:

- The contact surfaces should develop a frictional force that may pick up and hold the load with reasonably low pressure between the contact surfaces.

- The heat of friction should be rapidly dissipated and tendency to grab should be at a minimum.

- The surfaces should be backed by a material stiff enough to ensure a reasonably uniform distribution of pressure.

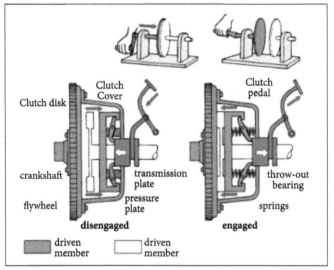

Clutch

The friction clutches of the following types are important from the subject point of view:

- Disc or plate clutches (single disc or multiple disc clutch).

- Cone clutches.

- Centrifugal clutches.

## Materials

Various materials have been used for the disc friction facings, including asbestos in the past. Modern clutches typically use a compound organic resin with copper wire facing or a ceramic material.

A typical coefficient of friction used on a friction disc surface is 0.35 for organic and 0.25 for ceramic. Ceramic materials are typically used in heavy applications such as trucks carrying large loads or racing, though the harder ceramic materials increase flywheel and pressure plate wear.

### 2.4.5 Transmission of Power by Single Plate, Multi Plate and Cone Clutches

### Single Plate Clutch

A single plate clutch is known as single disc clutch. It is shown in Figure below. It has two sides which are driving and the driven side. The driving side comprises of the driving shaft or engine crankshaft A. A boss B is keyed to it to which flywheel C is bolted as shown. On the driven side, there is a driven shaft D. It carries a boss E which can freely slide axially along with the driven shaft through splines F.

The clutch plate is mounted on the boss E. It is provided with rings of friction material – known as friction linings, on the both sides indicated H. One friction lining is pressed on the flywheel face and the other on the pressure plate I. A small spigot, bearing J, is provided in the end of the driving shaft for proper alignment.

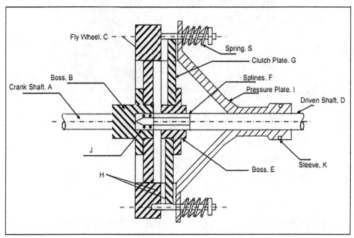

Single plate clutch

The pressure plate provides axial thrust or pressure between clutch plate G and the flywheel C and the pressure plate I through the linings on its either side, by means of the springs, S. The pressure plate remains engaged and as such clutch remains in operational position. Power from the driving shaft is transferred to the driven shaft from flywheel to the clutch plate through the friction lining between them.

From pressure plate the power is transmitted to clutch plate through friction linings. Both sides of the clutch plate are effective. When the clutch is to be disengaged the sleeve K is moved towards right hand side by means of clutch pedal mechanism (it is not shown in the figure). By doing this, there is no pressure between the pressure plate, flywheel and the clutch plate and no power is transmitted. In medium size and heavy vehicles, like truck, single plate clutch is used.

## Multi Plate Clutch

As already explained in a plate clutch, the torque is transmitted by friction between one or more pairs of co-axial annular faces kept in contact by an axial thrust provided by springs. In a single plate clutch, both sides of the plate are effective so that it has two pairs of surfaces in contact or n = 2.

Obviously, in a single plate clutch limited amount of torque can be transmitted. When large amount of torque is to be transmitted, more pair of contact surfaces is needed and it is precisely what is obtained by a multi-plate clutch.

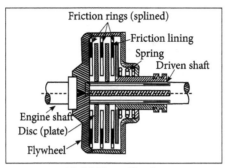

Multi Plate Clutch.

Cone clutches are friction clutches. They are simple in construction and are easy to disengage. However, the driving and driven shafts must be perfectly coaxial for efficient functioning of the clutch. This requirement is more critical for cone clutch compared to single plate friction clutch. A cone clutch consists of two working surfaces, viz. inner and outer cones, as shown in figure below.

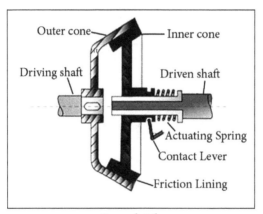

Cone clutch

The outer cone is fastened to the driving shaft and the inner cone is free to slide axially on the driven (output) shaft due to splines. A spring provides the necessary axial force to the inner cone to press against the outer cone, thus engaging the clutch.

A contact lever is used to disengage the clutch. The inner cone surface is lined with friction material. Due to wedging action between the conical working surfaces, there is considerable normal pressure and friction force with a small engaging force. The semi cone angle $\alpha$ is kept greater than a certain value to avoid self-engagement; otherwise disengagement of clutch would be difficult. This is kept around 12.50.

Let us prove that the torque transmitted by a cone clutch, when the intensity of pressure is uniform is given by,

$$T = \frac{2}{3} \frac{\mu W}{\sin \alpha} \left( \frac{r_1^3 - r_2^3}{r_1^2 - r_2^2} \right) \text{with usual notations.}$$

Let,

$P_n$=Intensity of pressure with which the conical friction surfaces are held together.

$r_1$, $r_2$=Outer and inner radius of frictions surfaces.

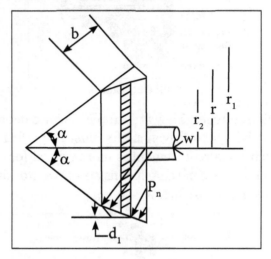

R = Mean radius of the friction surface = $\dfrac{r_1 \quad r_2}{}$

α = Semi cone angle

b = Face width (or) clutch face

Consider a small ring of radius r and thickness dr.

$d_1$ = Length of ring of the friction surface.

$\quad d_1 = dr\ \cosec\ \alpha$

$\quad$ Area of the ring $A = 2\pi r.\ d_1$

$\quad = 2\pi r.\ dr.\ \cosec\ \alpha$

Considering Uniform Pressure.

Normal load W= Normal pressure × Area of ring

$\quad = Pn \times 2\pi r\ dr\ \cosec\ \alpha$

Axial load $W = W_n \times \sin\ \alpha$

$\quad = P_n.\ 2\pi r\ dr.\ \cosec\ \alpha \sin\ \alpha \left[ \sin\ \alpha = \dfrac{1}{\cosec\ \alpha} \right]$

$\quad W = P_n.\ 2\pi r\ dr$

Total axial load transmitted to the clutch,

$$W = \int_{r_2}^{r_1} 2\pi \cdot P_n \cdot r \, dr = 2\pi P_n \cdot \int_{r_2}^{r_1} r \, dr$$

$$= 2\pi P_n \left[ \frac{r_1^2 - r_2^2}{2} \right]$$

$$W = \pi P_n \left[ r_1^2 - r_2^2 \right]$$

$$P_n = \frac{W}{\pi \left[ r_1^2 - r_2^2 \right]}$$

## Total Frictional Torque

$$T = \int_{r_2}^{r_1} 2\pi \mu P_n \, \text{Cosec} \, \alpha \, r^2 \, dr = 2\pi \mu P_n \cos ec \, \alpha \left[ \frac{r^3}{3} \right]_{r_2}^{r_1}$$

$$= 2\pi \mu P_n \cos ec \, \alpha \left[ \frac{r_1^3 - r_2^3}{3} \right]$$

$$= 2\pi \mu \cdot \frac{N}{\pi \left[ r_1^2 - r_2^2 \right]} \cos ec \, \alpha \left[ \frac{r_1^3 - r_2^3}{3} \right]$$

$$T = \frac{2}{3} \frac{\mu \cdot W}{\sin \alpha} \left[ \frac{r_1^3 - r_2^3}{r_1^2 - r_2^2} \right]$$

## Problem

1. A single plate friction clutch of both sides has 300 mm outer diameter and 160 mm inner diameter. The coefficient of friction is 0.2 and it runs at 1000 rpm. Let us determine the power transmitted for uniform wear and uniform pressure distributions cases if allowable maximum pressure is 0.08 Mpa.

Solution:

Given:

$N_1 = I = 2$, $D_0 = 300$ mm

$D_1 = 160$ mm $\mu = 0.2$

$N = 1000$ rpm $p = 0.08$ Mpa $= 0.08$ N /mm$^2$

## Uniform Wear Theory

Mean Diameter $D_m = \frac{D_0 - D_1}{2} = \frac{300 + 160}{2} = 230$mm

Axial Force:

$$Fa = \frac{1}{2}\pi b D_1 (D_o - D_1)$$

$$\therefore Fa = \frac{1}{2}\pi \times 0.08 \times 160(300 - 160)$$

$$= 2814.87 \text{ N}$$

Torque transmitted,

$$T = \frac{1}{2}\mu n^1 Fa D_m$$

$$T = \frac{1}{2} 0.2 \times 2 \times 2814.87 \times 230$$

$$T = 129484 \text{ N-mm}$$

Power transmitted,

$$P = \frac{2\pi MT}{60 \times 10^6}$$

$$P = \frac{2\pi \times 1000 \times 129484}{60 \times 10^6}$$

$$P = 13.56 \text{ kW}$$

## Uniform Wear Theory

$$D_m = \frac{2}{3} = \left(\frac{D_o^3 - D_1^3}{D_o^2 - D_1^2}\right)$$

Mean Diameter:

$$D_m = \frac{2}{3} = \left(\frac{300^3 - 160^3}{300^2 - 160^2}\right)$$

$$D_m = 237.1 \text{ mm}$$

Axial Force $Fa = \dfrac{\pi p (D_o^2 - D_1^2)}{4}$

$$Fa = \frac{\pi 0.08(300^2 - 160^2)}{4}$$

$$Fa = 4046. \text{ N}$$

Torque transmitted:

$$T = n^1 \frac{1}{2} \mu \, Fa \, D_m$$

$$T = 2\frac{1}{2} 0.2 \times 4046.4 \times 237.1$$

T=1918803.3 N-mm

Power transmitted:

$$P = \frac{2\pi nT}{60 \times 10^6}$$

$$P = \frac{2 \times \pi \times 1000 \times 191880.3}{60 \times 10^6}$$

P = 20.1 kW

2. A multi plate clutch of alternate bronze and steel plates is to transmit 6 kW power at 800 rpm. The inner radius is 38 mm and outer radius is 70 mm. The coefficient of friction is 0.1 and maximum allowable pressure is 350 kN /m². Let us determine:

   i.   Axial force required.

   ii.  Total number of discs.

   iii. Average pressure.

   iv.  Actual maximum pressure.

Solution:

Given:

   P = 60 kW, N = 800 rpm, $R_i$= 38mm,

   $D_i$= 76 mm,

   $R_o$= 70,

   $D_o$= 140mm, $\mu$= 0.1, P = 350 kN / m²= 0.35 N / mm²

## Axial Force

$$Fa = \frac{1}{2} \pi_1 p D_1 \left(D_o - D_1\right) - \left(13.32 \, DDH\right)$$

$$Fa = \frac{1}{2} \pi 0.35 \times 76 \left(146 - 76\right)$$

=2674.12 N

## Torque to be Transmitted

$$P = \frac{2\pi NT}{60 \times 10^6}$$

$$T = \frac{P \times 60 \times 10^6}{2\pi N} = \frac{6 \times 60 \times 10^6}{2 \times \pi \times 800} = 71625 N - mm$$

## Assuming Uniform Wear Theory

Mean Diameter $D_m = \frac{(D_2 + D_1)}{2} = \frac{140 + 76}{2} = 108\,mm,$

Torque transmitted,

$$T = \frac{1}{2} i\,\mu_1\, fa D_m$$

$$71625 = n\frac{1}{2} \times 0.1 \times 2674.12 \times 108$$

$n = 4.96 \cong 6$ (even number).

Number of driving (steel) discs $= n_1 = \frac{n}{2} = \frac{6}{2} = 3.$

Number of driving (bronze) discs $= n_2 = n_1 + 1 = 3 + 1 = 4.$

## Average Pressure Occurs at Mean Diameter

Axial force $Fa = \frac{1}{2}\pi_1 p D_m (D_0 - D_1)$

$$2674.12 = \frac{1}{2}\pi_1 p 108 (140 - 76)$$

$\therefore$ Average pressure $p = 0.246 N / mm^2$

## For 6 Friction Surface and Torque Transmitted

$$T = \frac{1}{2} i\,\mu_1\, Fa D_m$$

$$71625 = 6\frac{1}{2} 0.1_1\, Fa \times 108$$

$\therefore Fa = 2210.6 N$

Maximum pressure occur at inner radius:

$$\text{Axial force} = \frac{1}{2}\pi_1 p D_1 (D_o - D_i)$$

$$2210.6 = \frac{1}{2}\pi_1 \, p \, 76 (140 - 76)$$

Actual maximum pressure $P = 0.289 \text{ N/mm}^2$

# Brakes, Dynamometers and Gear Trains

**3**

## 3.1 Brakes and its Classification

**On the Basis of Method of Actuation**

- Foot brake (also called service brake) operated by foot pedal.
- Hand brake – it is also called parking brake operated by hand.

**On the Basis of Mode of Operation**

- Mechanical brakes.
- Hydraulic brakes.
- Air brakes.
- Vacuum brakes.
- Electric brakes.

**On the Basis of Action on Front or Rear Wheels**

- Front-wheel brakes.
- Rear-wheel brakes.

**On the Basis of Method of Application of Braking Contact**

- Internally – expanding brakes.
- Externally – contracting brakes.

### 3.1.1 Analysis of Simple Block

Frictional brakes are most common and can be divided broadly into "shoe" or "pad" brakes, using an explicit wear surface and hydrodynamic brakes, such as parachutes, which use friction in a working fluid and do not explicitly wear. Typically the term "friction brake" is used to mean pad/shoe brakes and excludes hydrodynamic brakes, even though hydrodynamic brakes use friction.

## Clutch

Clutch is a transmission device of an Automobile which is used to engage and disengage the power from the engine to the rest of the system.

## Brake

A brake is a device by means of which artificial resistance is applied to a moving machine member, in order to retard or stop the motion of a machine.

A single block or shoe brake is shown in figure. It consists of a block or shoe which is pressed against the rim of a revolving brake wheel drum. The block is made of a softer material than the rim of the wheel. This type of a brake is commonly used on railway trains and tram cars.

The friction between the block and the wheel causes a tangential braking force to act on the wheel, which retard the rotation of the wheel. The block is pressed against the wheel by a force applied to one end of a lever to which the block is rigidly fixed as shown in Figure. The other end of the lever is pivoted on a fixed fulcrum O.

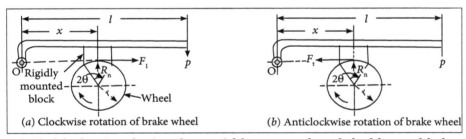

(a) Clockwise rotation of brake wheel     (b) Anticlockwise rotation of brake wheel

Single block brake. Line of action of tangential force passes through the fulcrum of the lever.

Let,

$P$ = Force applied at the end of the lever,

$r$ = Radius of the wheel,

$R_N$ = Normal force pressing the brake block on the wheel,

$2\theta$ = Angle of contact surface of the block,

$\mu$ = Coefficient of friction,

$F_t$ = Tangential braking force or the frictional force acting at the contact surface of the block and the wheel.

If the angle of contact is less than $60°$, then it may be assumed that the normal pressure between the block and the wheel is uniform. In such cases, tangential braking force on the wheel,

$$F_t = \mu.R_N \qquad\qquad ...(i)$$

and the braking torque, $T_B = F_t.r = \mu.R_N.r$           ... (ii)

Let us now consider the following three cases:

Case 1: When the line of action of tangential braking force ($F_t$) passes through the fulcrum O of the lever, and the brake wheel rotates clockwise as shown in Figure (a) then for equilibrium, taking moments about the fulcrum O, we have,

$$R_N \times x = P \times l \text{ Or } R_N = \frac{P \times l}{x}$$

∴ Braking torque,

$$T_B = \mu.R_N.r = \mu \times \frac{P.l}{x} \times r = \frac{\mu.P.l.r}{x}$$

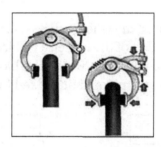

When brakes are on, the pads grip the wheel rim from either side, friction between the pads and the rim converts the cycle's kinetic energy into heat as they reduce its speed.

It may be noted that when the brake wheel rotates anticlockwise as shown in Figure (b), then the braking torque is same, i.e.

$$T_B = \mu.R_N.r = \frac{\mu.P.l.r}{x}$$

Case 2: When the line of action of the tangential braking force ($F_t$) passes through a distance 'a' below the fulcrum O, and the brake wheel rotates clockwise as shown in Figure(a), then for equilibrium, taking moments about the fulcrum O,

$$R_N \times x + F_t \times a = P.l$$

$$R_N \times x + \mu R_N \times a = P.l$$

$$R_N = \frac{P.l}{x + \mu.a}$$

and braking torque,

$$T_B = \mu R_N.r = \frac{\mu.p.l.r}{x + \mu.a}$$

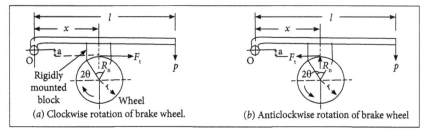

Single block brake. Line of action of Ft passes below the fulcrum.

When the brake wheel rotates anticlockwise, as shown in Figure (b), then for equilibrium,

$$R_N.x = P.l + F_t.a = P.l + \mu.R_N.a$$

$$\text{Or } R_N(x - \mu.a) = P.l \text{ or } R_N = \frac{P.l}{x - \mu.a}$$

and braking torque, $T_B = \mu.R_N.r = \dfrac{\mu.P.l.r}{x - \mu.a}$

Case 3: When the line of action of the tangential braking force ($F_t$) passes through a distance 'a' above the fulcrum O, and the brake wheel rotates clockwise as shown in Figure (a), then for equilibrium, taking moments about the fulcrum O, we have,

$$R_N.x = P.l + F_t.a = P.l + \mu.R_N.a$$

$$\text{Or } R_N(x - \mu.a) = P.l \text{ or } R_N = \frac{P.l}{x - \mu.a}$$

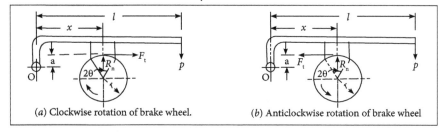

(a) Clockwise rotation of brake wheel.  (b) Anticlockwise rotation of brake wheel

Single block brake. Line of action of $F_t$ passes above the fulcrumand braking torque,

$$T_B = \mu.R_N.r = \frac{\mu.P.l.r}{x - \mu.a}$$

When the brake wheel rotates anticlockwise as shown in Figure (b), then for equilibrium, taking moments about the fulcrum O, we have,

$$R_N \times x + F_t \times a = P.l \text{ or } R_N \times x + \mu.R_N \times a = P.l \text{ or } R_N = \frac{P.l}{x + \mu.a}$$

and braking torque, $T_B = \mu.R_N.r = \dfrac{\mu.P.l.r}{x + \mu.a}$

## 3.1.2 Band and Internal Expanding Shoe Brake: Braking of a Vehicle

### Band Brake

A band brake consists of a flexible band of leather, one or more ropes, or steel lined with friction material, which embraces a part of a circumference of the drum. A band brake is called a simple band brake in which one end of the band is attached to a fixed pin or fulcrum of the lever while the other end is attached to the lever at a distance be from the fulcrum.

When a force P is applied to the lever at C, the lever turns about the fulcrum pin 'O' and tightens the band on the drum and hence the brakes are applied. The friction between the band and the drum provides the braking force.

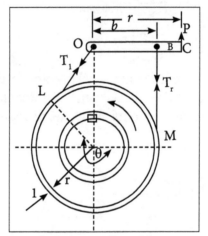

Clockwise rotation of drum.

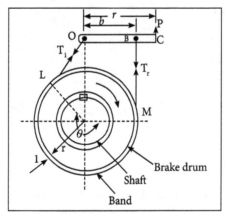

Anticlockwise rotation of drum.

Let,

$T_1$ = Tension in the tight side of the band.

$T_2$ = Tension in the slack side.

$\theta$ = Angle of lap.

$\mu$ = Coefficient of friction between the band and the drum.

$\gamma$ = Radius of the drum.

t = Thickness of the band.

$\gamma_e$ = Effective radius of the drum = $\gamma + \dfrac{t}{2}$.

Tension ratio is,

$$\frac{T_1}{T_2} = e^{\mu\theta}$$

Braking torque on the drum,

$$T_B = (T_1 - T_2)\gamma$$

Now taking moments about the fulcrum O,

$$P.l = T_1\, b$$

$$P.l = T_2\, b$$

If the permissible tensile stress ($\sigma$) for the material of the band is known, then maximum tension in the band is given by,

$$T_1 = \sigma\, Wt$$

## Internal Expanding Shoe Brake

An internal expanding brake consists of two shoes $S_1$ and $S_2$ as shown in Figure. The outer surface of the shoes are lined with some friction material to increase the coefficient of friction and to prevent wearing away of the metal.

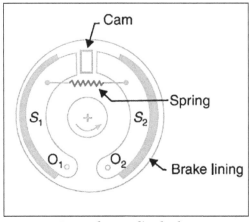

Internal expanding brake.

Both shoe is pivoted at one end about a fixed fulcrum $O_1$ and $O_2$ and made to contact a cam at the other end. When the cam rotates, the shoes are pushed outwards against the rim of the drum.

The friction between the shoes and the drum produces the braking torque and hence reduces the speed of the drum. The shoes are normally held in off position by a spring as shown in the figure. The drum encloses the entire mechanism to keep out dust and moisture. This type of brake is commonly used in motor cars and light trucks.

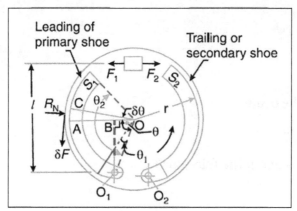

Forces on an internal expanding brake.

We shall now consider the forces acting on such a brake, when the drum rotates in the anticlockwise direction as shown in figure. It may be noted that for the anticlockwise direction, the left hand shoe is known as leading or primary shoe while the right hand shoe is known as trailing or secondary shoe.

Let,

   $r$ = Internal radius of the wheel rim,

   $b$ = Width of the brake lining,

   $p_1$ = Maximum intensity of normal pressure,

   $p_N$ = Normal pressure,

   $F_1$ = Force exerted by the cam on the leading shoe,

   $F_2$ = Force exerted by the cam on the trailing shoe.

Consider a small element of the brake lining AC subtending an angle $\delta\theta$ at the centre. Let OA makes an angle $\theta$ with $OO_1$ as shown in Figure. It is assumed that the pressure distribution on the shoe is nearly uniform, however the friction lining wears out more at the free end.

Since the shoe turns about $O_1$, therefore the rate of wear of the shoe lining at A will be proportional to the radial displacement of that point. The rate of wear of the shoe lining varies directly as the perpendicular distance from $O_1$ to OA, i.e. $O_1B$. From the geometry of the figure,

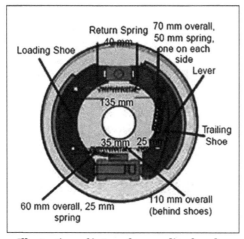

Illustration of internal expanding breaks.

$$O_1B = OO_1 \sin\theta$$

and normal pressure at A,

$$p_N \propto \sin\theta \text{ or } p_N = p_1 \sin\theta$$

∴ Normal force acting on the element,

$\delta R_N$ = Normal pressure × Area of the element

$$= p_N(b.r. \delta\theta) = p_1 \sin\theta(b.r. \delta\theta)$$

and braking or friction force on the element,

$$\delta F = \mu \times \delta R_N = \mu.p_1 \sin\theta(b.r. \delta\theta)$$

∴ Braking torque due to the element about O,

$$\delta T_B = \delta F \times r = \mu.p_1 \sin\theta(b.r. \delta\theta)r = \mu. p_1 b r^2 (\sin\theta.\delta\theta)$$

and total braking torque about *O* for whole of one shoe,

$$T_B = \mu p_1 b r^2 \int_{\theta_1}^{\theta_2} \sin\theta d\theta = \mu p_1 b r^2 \left[-\cos\theta\right]_{\theta_1}^{\theta_2}$$

$$= \mu p_1 b r^2 (\cos\theta_1 - \cos\theta_2)$$

$$\delta M_N = \delta R_N \times O_1B = \delta R_N (OO_1 \sin\theta)$$

$$= p_1 \sin\theta(b.r.\delta\theta)(OO_1\sin\theta) = p_1 \sin^2\theta(b.r.\delta\theta)OO_1$$

Total moment of normal forces about the fulcrum $O_1$,

$$M_N = \int_{\theta_1}^{\theta_2} p_1 \sin^2\theta (b.r.\delta\theta) OO_1 = p_1 b.r.OO_1 \int_{\theta_1}^{\theta_2} \sin^2\theta \, d\theta$$

$$= P_1.b.r.OO_1 \int_{\theta_1}^{\theta_2} \frac{1}{2}(1 - \cos 2\theta) \, d\theta \quad \ldots \left[ \therefore \sin^2\theta = \frac{1}{2}(1 - \cos 2\theta) \right]$$

$$= \frac{1}{2} P_1.b.r.OO_1 \left[ \theta - \frac{\sin 2\theta}{2} \right]_{\theta_1}^{\theta_2}$$

$$= \frac{1}{2} P_1.b.r.OO_1 \left[ \theta_2 - \frac{\sin 2\theta_2}{2} - \theta_1 + \frac{\sin 2\theta_1}{2} \right]$$

$$= \frac{1}{2} P_1.b.r.OO_1 \left[ \theta_2 - \frac{\sin 2\theta_2}{2} - \theta_1 + \frac{\sin 2\theta_1}{2} \right]$$

Moment of frictional force $\delta F$ about the fulcrum $O_1$,

$$\delta M_F = \delta F \times AB = \delta F(r - OO_1 \cos\theta) \ldots (AB = r - OO_1 \cos\theta)$$

$$= \mu p_1 \sin\theta \, (b.r.\delta\theta)(r - OO_1 \cos\theta)$$

$$= \mu.p_1.b.r \, (r\sin\theta - OO_1 \sin\theta\cos\theta)$$

$$= \mu.p_1.b.r \left( r\sin\theta - \frac{OO_1}{2} \sin 2\theta \right) \delta\theta \qquad \ldots (\because 2\sin\theta\cos\theta = \sin\theta)$$

$\therefore$ Total moment of frictional force about the fulcrum $O_1$,

$$M_F = \mu p_1 br \int_{\theta_1}^{\theta_2} \left( r\sin\theta - \frac{OO_1}{2} \sin 2\theta \right) d\theta$$

$$= \mu p_1 br \left[ -r\cos\theta + \frac{OO_1}{4} \cos 2\theta \right]_{\theta_1}^{\theta_2}$$

$$= \mu p_1 br \left[ -r\cos\theta_2 + \frac{OO_1}{4} \cos 2\theta_2 + r\cos\theta_1 - \frac{OO_1}{4} \cos 2\theta_1 \right]_{\theta_1}^{\theta_2}$$

$$= \mu p_1 br \left[ -r(\cos\theta_1 - \cos\theta_2) + \frac{OO_1}{4}(\cos 2\theta_2 - \cos 2\theta_1) \right]$$

Now for leading shoe, taking moments about the fulcrum $O_1$,

$$F_1 \times l = M_N - M_F$$

and for trailing shoe, taking moments about the fulcrum $O_2$,

$$F_2 \times l = M_N + M_F$$

## Problems

1. A simple brake as shown in the figure is used on a shaft carrying a flywheel of mass 450 kg. The radius of gyration of the flywheel is 500 mm. and runs at 320 rpm. The co-efficient of friction is 0.2 and the diameter of brake drum is 250 mm. Let us determine the following:

- Torque applied due to a hand load of 150 N.

- The number of turns of the wheel before it is brought to rest.

- The time required to bring it to rest from the moment of application of the brake.

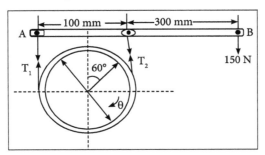

Solution:

Given data:

$\qquad$ m = 450 kg

$\qquad$ K = 500 mm= 0.5 m

$\qquad$ N = 320 rpm

$\qquad$ μ = 0.2

$\qquad$ a = 100 mm = 0.1 m

$\qquad$ l = 300 mm = 0.3 m and β= 60°

$\qquad$ D = 250 mm = > r = 125 mm = 0.125 m

### Brake Torque Applied at the Lever End

Angle of contact,

$\qquad$ θ = 360° - β

$\qquad$ = 360° - 60° $\Rightarrow$ θ = 300°

$\qquad$ $\theta = 300 \times \dfrac{\pi}{180}$

$\qquad$ θ = 5.235 rad

Tension ratio is given by,

$$\frac{T_1}{T_2} = e^{\mu\theta}$$

$$\frac{T_1}{T_2} = e^{0.2 \times 5.235}$$

$$T_1 = 2.849\,T_2 \qquad\qquad\qquad\qquad ...(1)$$

Taking Moments about the fulcrum o, we get,

$$F \times l = T_1 \times a$$

$$150 \times 0.3 = T_1 \times 0.1$$

$$T_1 = 450\ N$$

Substituting $T_1 = 450$ in equation (1),

$$T_2 = \frac{T_1}{2.849} = \frac{450}{2.849}$$

$$T_2 = 157.95\,N$$

Braking torque is given by,

$$T_B = (T_1 - T_2)r$$

$$T_b = (450 - 157.95) \times 0.125$$

$$T_b = 36.506\ Nm$$

## Number of Turns of Flywheel Before it Comes to Rest (n)

Kinetic Energy of Flywheel,

$$K.E = \frac{1}{2}I\omega^2$$

$$= \frac{1}{2} \times 450 \times 0.5^2 \times \left[\frac{2\pi \times 320}{60}\right]^2 \quad \because I = mk^2$$

$$K.E = 63.165\,KNm$$

Kinetic Energy is used to overcome the work done due to the braking torque ($T_B$).

$$K.E\ of\ flywheel = T_B \times \omega$$

$$63.165 \times 10^3 = 36.506 \times 2\pi n$$

$$n = 275\ revolutions$$

Time taken by the flywheel to come to rest:

$$Time\ taken = \frac{n}{N} = \frac{275}{320} = 0.859\,minute$$

Result:

Applied Torque $T_B$ = 36.506 Nm

Number of trans of flywheel = n = 275 revolutions

Time taken = 0.859 Minutes

2. A rope drive is to transmit 250 kW from a pulley of 1.2 m diameter, running at a speed of 300 r.p.m. The angle of lap may be taken as $\pi$radians. The groove half angle is 22.5°. The ropes to be used are 50 mm in diameter. The mass of the rope is 1.3 kg per metre length and each rope has a maximum pull of 2.2 kN, the coefficient of friction between rope and pulley is 0.3. Let us determine the number of ropes required. If the overhang of the pulley is 0.5 m, suggest suitable size for the pulley shaft if it is made of steel with a shear stress of 40 MPa.

Solution:

Given:

P= 250 kW = 250 × 10³W

d= 1.2 m; N= 300 r.p.m; $\theta$= $\pi$rad; $\beta$= 22.5°

dr= 50 mm ; m= 1.3 kg /m

T= 2.2 kN = 2200 N

$\mu$= 0.3; $\tau$= 40 MPa = 40 N/mm²

To find:

The number of ropes.

Formula to be used:

We know that the velocity of belt,

$$v=\frac{\pi d \cdot N}{60}=\frac{\pi \times 1.2 \times 300}{60}=18.85 \text{ m/s}$$

and centrifugal tension, $T_C$ =m.v² = 1.3 (18.85)²= 462 N

∴ Tension in the tight side of the rope,

$$T_1 = T - T_C = 2200 - 462 = 1738 \text{ N}$$

Let $T_2$ = Tension in the slack side of the rope.

We know that,

$$2.3 \log\left(\frac{T_1}{T_2}\right) = \mu \cdot \theta \cdot \cos ec\,\beta = 0.3 \times \pi \times \cos ec\,22.5° = 0.9426 \times 2.6131 = 2.46.3$$

$$\log\left(\frac{T_1}{T_2}\right) = \frac{2.463}{2.3} = 1.071 \text{ or } \frac{T_1}{T_2} = 11.8 \text{ ...(Taking antilog of 1.071)}$$

$$T_2 = \frac{T_1}{11.8} = \frac{1738}{11.8} = 147.3 N$$

## Number of Ropes Required

We know that power transmitted per rope,

$$= (T_1 - T_2)\, v = (1738 - 147.3) \times 18.85 = 29\,985\ W = 29.985\ kW$$

∴ Number of ropes required

$$= \frac{\text{Total power transmitted}}{\text{Power transmitted per rope}} = \frac{250}{29.985} = 8.34 \text{ say } 9$$

## Diameter for the Pulley Shaft

Let,

D= Diameter for the pulley shaft.

We know that the torque transmitted by the pulley shaft,

$$T = \frac{P \times 60}{2\pi N} = \frac{250 \times 10^3 \times 60}{2\pi \times 300} = 7957\ N\text{-}m$$

Since the overhang of the pulley is 0.5 m, therefore bending moment on the shaft due to the rope pull,

$$M = (T_1 + T_2 + 2T_C)\,0.5 \times 9 \quad (\because \text{No of ropes} = 9)$$
$$= (1738 + 147.3 + 2 \times 462)\,0.5 \times 9 = 12642\ N\text{-}m$$

Equivalent twisting moment,

$$\dot{T}_e = \sqrt{T^2 + M^2} = \sqrt{(7957)^2 + (12642)^2} = 14938\ N\text{-}m$$
$$= 14.938 \times 10^6\ N\text{-}mm$$

We know that the equivalent twisting moment $(T_e)$,

$$= 14.938 \times 10^6 = \frac{\pi}{16} \times \tau \times D^3 = \frac{\pi}{16} \times 40 \times D^3 = 7.855\,D^3$$

3. A workshop crane is lifting a load of 25 kN through a wire rope and a hook. The weight of the hook etc. is 15 kN. The rope drum diameter may be taken as 30 times the diameter of the rope. The load is to be lifted with an acceleration of 1 m/s². Let us calculate the diameter of the wire rope. Take a factor of safety of 6 and Young's modulus for the wire rope 80 kN/mm². The ultimate stress may be taken as 1800 MPa. The cross-sectional area of the wire rope may be taken as 0.38 times the square of the wire rope diameter.

Solution:

Given:

$W = 25$ kN $= 25\,000$ N

$w = 15$ kN $= 15\,000$ N

$D = 30$ d; $a = 1$ m/s²

$E_r = 80$ kN/mm² $= 80 \times 10^3$ N/mm²

$\sigma_u = 1800$ MPa $= 1800$ N/mm²

$A = 0.38$ d

To find:

The diameter of the wire rope.

d= Diameter of wire rope in mm.

We know that direct load on the wire rope,

$$W_d = W + w = 25000 + 15000 = 40000 \text{ N}$$

Let us assume that a 6 × 19 wire rope is used. Therefore from above the table. We find that the diameter of wire,

$$d_w = 0.063 \text{ d}$$

We know that bending load on the rope,

$$W_b = \frac{E_r \times d_w}{D} \times A = \frac{80 \times 10^3 \times 0.63d}{30d} \times 0.38d^2 = 63.84d^2 \text{ N}$$

and load on the rope due to acceleration,

$$W_a = \frac{W + w}{g} \times a = \frac{25000 + 15000}{9.81} \times 1 = 4080 \text{ N}$$

Total load acting on the rope,

$$= W_d + W_b + W_a = 40000 + 63.84d^2 + 4080$$
$$= 44080 + 63.84d^2 \qquad\qquad\qquad ...(i)$$

We know that total load on the rope,

= Area of wire rope × Allowable stress

$$= A \times \frac{\sigma_u}{F.S.} = 0.38d^2 \times \frac{1800}{6} = 114d^2 \qquad \text{...(ii)}$$

From equations (i) and (ii), we have,

$$44080 + 63.84d^2 = 114d^2$$

$$d^2 = \frac{44080}{114 - 63.84} = 879 \text{ or } d = 29.6\text{mm}.$$

From Table, we find that standard nominal diameter of 6 × 19 wire rope is 32 mm.

## 3.2   Dynamometer: Absorption and Transmission

Dynamometer is a device by means of which energy or work done by a prime mover can be measured. A mechanical dynamometer is essentially a brake with additional device to reduce frictional resistance by allowing the prime mover to run at the rated speed. Dynamometers are mainly classified into two types Absorption dynamometers and Transmission dynamometers.

Absorption Dynamometers: In an absorption dynamometer, the work done or energy of the prime mover is converted into heat usually against friction being measured. In other it is a braking system in which some provision is made for measuring friction torque on the drum. Examples are prony brake, rope brake, hydraulic dynamometers. These dynamometers can be used to measure moderate power produced by various types of prime movers. A brief description of various absorption dynamometers is given further.

Transmission Dynamometers, the energy is not wasted in friction but is used for doing work. The energy or power produced by the engine is transmitted through the dynamometer to some other machines where the power developed is suitably measured.

### 3.2.1 Prony Brake

A prony brake dynamometer is the simplest form of an absorption type dynamometer as shown in the figure. It consists of two wooden blocks placed around a pulley fixed to the shaft of an engine whose power is required to be measured. The blocks are clamped by using two bolts and nuts, as shown in the figure.

A helical spring is provided between the nut and upper block to adjust the pressure on the pulley to control its speed. The upper block has a long lever attached to it and it carries a weight $W$ at its outer end. A counter weight is placed at the other end of the lever which balances the brake when unloaded. Two stops S, S are provided to limit the motion of the lever.

When brake is to be put in operation, the long end of the lever is loaded with suitable weights W and the nuts are tightened until the engine shaft runs at a constant speed and lever is in horizontal position. Under these conditions, the moment due to the weight W must balance the moment of the frictional resistance between the blocks and the pulley.

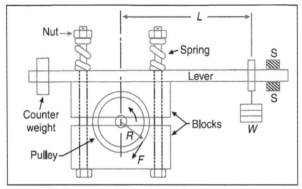

Prony brake dynamometer.

Let,

W = Weight at the outer end of the lever in newtons,

L = Horizontal distance of the weight W from the centre of the pulley in metres,

F = Frictional resistance between the blocks and the pulley in newtons,

R= Radius of the pulley in metres,

N = Speed of the shaft in r.p.m.

We know that the moment of the frictional resistance or torque on the shaft,

T = W.L = F.R N-m

Work done in one revolution:

= Torque × Angle turned in radians

= T × 2π N-m

∴ Work done per minute

= T ×2π N N-m

We know that brake power of the engine,

$$\text{B.P} = \frac{\text{Work done per min}}{60} = \frac{T \times 2\pi N}{60} = \frac{W.L \times 2\pi N}{60} \text{ Watts}$$

## 3.2.2 Rope Brake

Another form of absorption type dynamometer which is most commonly used for measuring the brake power of the engine is the Rope brake dynamometers. It consists of one, two or more ropes wound around the flywheel or rim of a pulley fixed rigidly to the shaft of an engine.

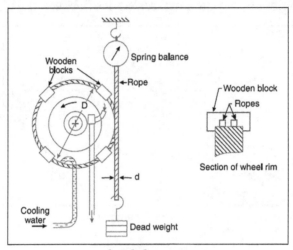

Rope break dynamometer.

The upper end of ropes is attached to a spring balance while the lower end of the ropes is kept in position by applying a dead weight as shown in figure. In order to prevent the slipping of the rope over the flywheel, wooden blocks are placed at intervals around the circumference of the flywheel.

In the operation of the brake, engine is made to run at a constant speed. The frictional torque, due to rope, must be equal to the torque being transmitted by the engine.

Let,

W = Dead load in newtons,

S = Spring balance reading in newtons,

D = Diameter of the wheel in metres,

d = diameter of rope in metres, and

N = Speed of the engine shaft in r.p.m.

∴ Net load on the brake,

= (W – S) N

We know that distance moved in one revolution,

= π(D+ d)m

∴ Work done per revolution,

$$= (W - S)\,\pi\,(D + d)\,\text{N-m}$$

and work done per minute,

$$= (W - S)\,\pi(D + d)\,N\ \text{N-m}$$

∴ Brake power of the engine,

$$\text{B.P} = \frac{\text{Work done per min}}{60} = \frac{(W-S)\pi(D+d)N}{60}\text{watts}$$

If the diameter of the rope ($d$) is neglected, then brake power of the engine,

$$\text{B.P} = \frac{(W-S)\pi D N}{60}\text{watts}$$

## 3.2.3 Band Brake Dynamometer

Band brake dynamometer is shown in Figure, It can be can considered as a for of rope brake dynamometer. In this, instead of rope having two ends. here. an integrated belt with three ends is used. The two lower ends, one on either side of the brake drum carry loads separately and the third upper end is connected to the spring balance.

Let,

W = Weight on right hand side (N)

w= Weight on left hand side (N)

S = Spring balance reading (N)

$r_e$ = effective radius of the brake drum (m)

N = speed in r.p.m

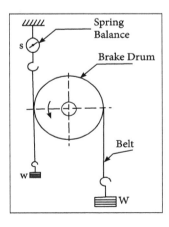

Then,

Braking Torque, $T = [(W+w)-S] \times r_e$

$$\text{Brake Power} = T \times \frac{2\pi N}{60}$$

$$= [W + w - S] \times r_e \times \frac{2\pi N}{60}$$

### 3.2.4 Belt Transmission Dynamometer

A belt transmission dynamometer measures the difference between the tensions on the tight and slack sides of a belt when it is moving from one pulley to another pulley.

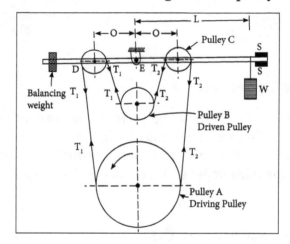

This difference of tensions (i.e. $T_1$-$T_2$) when multiplied by the speed of the belt, gives the power transmitted. Figure show a belt transmission dynamometer which is also called Tat-ham dynamometer. It consists of a driving pulley A, driven pulley B and intermediate pulleys C and D.

The driving pulley A is rigidly fixed to the shaft of an engine whose power is to be measured. The driven pulley B is fixed to another shaft to which power is to be transmitted. The intermediate pulleys C and D rotates pins fixed to the lever.

The lever is pivoted at E, the mid point of the two intermediate pulley centers. A continuous belt runs over the driving and the driven pulleys through the two intermediate pulleys. The movement of lever is controlled between two stops S and S one on each side of the lever.

Let the driving pulley A rotates anti-clockwise. The tight and slack sides of the belt will be as shown in figure. The total downward force acting on pulley D is $2T_1$, whereas the total downward force on pulley C is $2T_2$. As $2T_1$ is greater than $2T_2$ therefore the lever starts rotating about E in anti-clockwise direction. In order to balance it, a weight W is suspended at a distance L from E on the lever as shown in figure.

When the lever is in horizontal position, the total moments of all the flange about fulcrum E should be given i.e.

Total anti-Clockwise moment = Total clockwise moment

$$2T_1 \times a = 2T_2 \times a + W \times L$$

$$2T_1 \times a - 2T_2 \times a = W \times L$$

$$2a(T_1 - T_2) = W \times L$$

$$(T_1 - T_2) = \frac{W \times L}{2a}$$

Let,

v=Belt speed in m/s,

D=Dia. of pulley A,

N=Speed of engine shaft,

Then,

$$v = \frac{\pi DN}{60}$$

Power of the engine,

$$= (T_1 - T_2) \times v$$

$$= (T_1 - T_2) \times \frac{\pi DN}{60} \text{ watts}$$

### 3.2.5 Torsion Dynamometer

When the belt is transmitting power from one pulley to another, the tangential effort on the driven pulley is equal to the difference between the tensions in the tight and slack sides of the belt. A belt dynamometer is introduced to directly measure the difference between the tensions of the belt, while it is running.

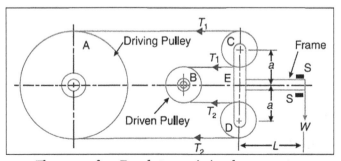

Throneycroft or Froude transmission dynamometer.

A belt transmission dynamometer, as shown in Figure, is called a Frode or Throney-croft transmission dynamometer. It consists of a pulley A (called driving pulley) which is rigidly fixed to the shaft of an engine whose power is required to be measured.

There is another pulley B (called driven pulley) mounted on another shaft to which the power from the pulley A is transmitted. The pulleys A and B are connected by means of a continuous belt passing round the two loose pulleys C and D which are mounted on a T-shaped frame.

The frame is pivoted at E and its movement is controlled by the two stops S,S. Since the tension in the tight side of the belt ($T_1$) is greater than tension in the slack side of the belt ($T_2$), therefore the total force acting on the pulley C (i.e. $2T_1$) is greater than the total force acting on the pulley D (i.e. $2T_2$). It is thus obvious that the frame causes movement about E in the anticlockwise direction. In order to balance it, a weight W is applied at a distance L from E on the frame as shown in Figure.

Now taking moments about the pivot E, neglecting friction,

$$2T_1 \times a = 2T_2 \times a + W.L \text{ or } T_1 - T_2 = \frac{W.L}{2a}$$

Let,

D = diameter of the pulley A in metres

N = Speed of the engine shaft in r.p.m

$\therefore$ Work done in one revolution = $(T_1 - T_2)\pi D$ N-m

and work done per minute = $(T_1 - T_2)\pi DN$ N-m

$\therefore$ Brake power of the engine,

$$B.P. = \frac{(T_1 - T_2)\pi DN}{60} \text{ watts}$$

## 3.3   Gear Trains

A gear train is said to be simple if the shaft has only one gear mounted on it. The typical spur gears are shown in the figure below. The direction of rotation is reversed from one gear to another. It has no effect on the gear ratio. The teeth on the gears must be of the same size. If gear A advances one tooth, so does B and C.

t= Number of teeth on the gear

D= Pitch circle diameter, N= Speed in rpm

m= Module =D/t

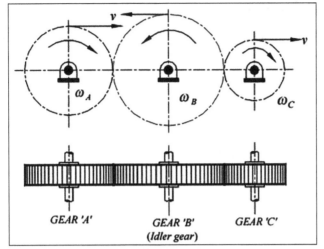

Simple gear train

Module must be same for all gears otherwise they would not mesh,

$$m = \frac{D_A}{t_A} = \frac{D_B}{t_B} = \frac{D_C}{t_C}$$

$$D_A = m\,t_A;\; D_B = m\,t_B \text{ and } D_C = m\,t_C$$

$\omega$=angular velocity,

v=linear velocity on the circle,

$$v = \omega\frac{D}{2} = \omega r$$

The velocity v of any point on the circle must be the same for all the gears, otherwise they would be slipping,

$$v = \omega_A\frac{D_A}{2} = \omega_B\frac{D_B}{2} = \omega_C\frac{D_C}{2}$$

$$\omega_A D_A = \omega_B D_B = \omega_C D_C$$

$$\omega_A m t_A = \omega_B m t_B = \omega_C m t_C$$

$$\omega_A t_A = \omega_B t_B = \omega_C t_C$$

Or in terms of rev/min,

$$NA_tA = N B_t B = NC_tC$$

## Application

- To connect gears where a large center distance is required.

- To obtain high speed ratio.

- To obtain desired direction of motion of the driven gear.

### 3.3.1 Simple and Compound Train

## Simple Gear Train

The simple gear train is used where there is a large distance to be covered between the input shaft and the output shaft. Each gear in a simple gear train is mounted on its own shaft.

When examining simple gear trains, it is necessary to decide whether the output gear will turn faster, slower, or the same speed as the input gear. The circumference of these two gears will determine their relative speeds.

Suppose the input gear's circumference is larger than the output gear's circumference. The output gear will turn faster than the input gear. On the other hand, the input gear's circumference could be smaller than the output gear's circumference. In this case the output gear would turn more slowly than the input gear. If the input and output gears are exactly the same size, they will turn at the same speed.

In many simple gear trains there are several gears between the input gear and the output gear.

These middle gears are called idler gears. Idler gears do not affect the speed of the output gear.

## Compound Gear Train

In a compound gear train at least one of the shafts in the train must hold two gears.

Compound gear trains are used when large changes in speed or power output are needed and there is only a small space between the input and output shafts.

The number of shafts and direction of rotation of the input gear determine the direction of rotation of the output gear in a compound gear train. The train in Figure has two gears in between the input and output gears. These two gears are on one shaft. They rotate in the same direction and act like one gear. There are an odd number of gear shafts in this example. As a result, the input gear and output gear rotate in the same direction.

Since two pairs of gears are involved, their ratios are compounded or multiplied together.

## Sub-Classification of Compound Gear Trains

Compound gear trains are further sub-divided into two classes:

- When the axes of the first gear (i.e. first driver) and the last gear (i.e. last driven or follower) are co-axial, then the gear train is known as reverted gear train. In a reverted gear train, the motion of the first gear and the last gear is like.

- Non-reverted gear train: Any compound gear train in which first and last gears are not co-axial, are called non-reverted type gear train.

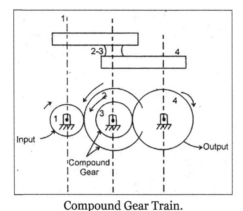

Compound Gear Train.

## Gear Ratio

The gear ratio of a gear train, also known as its speed ratio, is the ratio of the angular velocity of the input gear to the angular velocity of the output gear. The gear ratio can be calculated directly from the number of teeth on the gears in the gear train.

$$R = \frac{\omega_A}{\omega_B} = \frac{N_B}{N_A}$$

Where,

$\omega_A$, $\omega_B$- Angular velocity of input and output gear respectively,

$N_a$, $N_b$- Number of teeth on the input gear and output gear respectively.

### 3.3.2 Reverted Train

A reverted gear train is a compound gear train in which, the first and last gears are co-axial. They find applications in clocks and in simple lathes where back gear is used to impact slow speed to the chuck.

When the axes of the first gear (i.e. first driver) and the last gear (i.e. last driven or follower) are co-axial, then the gear train is known as reverted gear train as shown in Figure. We see that gear 1 (i.e. first driver) drives the gear 2 in the opposite direction.

Since the gears 2 and 3 are mounted on the same shaft, therefore they form a compound gear and the gear 3 will rotate in the same direction as that of gear 2. The gear 3 (which is now the second driver) drives the gear 4 (i.e. the last driven or follower) in the same direction as that of gear 1.Thus we see in a reverted gear train, the motion of the first gear and last gear is like.

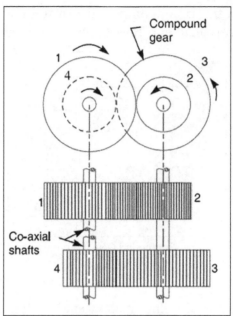

Reverted Gear Train.

Let,

$T_1$ = Number of teeth on gear 1.

$r_1$ = Pitch circle radius of gear 1.

$N_1$= Speed of gear 1 in r.p.m,

Similarly,

$T_2, T_3, T_4$ = Number of teeth on respective gears,

$r_2, r_3, r_4$ = Pitch circle radii of respective gears,

$N_2, N_3, N_4$ = Speed of respective gears in r.p.m,

Since the distance between the centres of the shafts of gears 1 and 2 as well as gears 3 and 4 is same, therefore,

$$r_1 + r_2 = r_3 + r_4 \qquad \qquad \text{...(1)}$$

Also, the circular pitch or module of all the gears is assumed to be same, therefore number of teeth on each gear is directly proportional to its circumference or radius.

$$T_1 + T_2 = T_3 + T_4 \qquad \qquad \text{...(2)}$$

$$\text{speed ratio} = \frac{\text{Product of number of teeth on drivens}}{\text{Product of number of teeth on drivers}} \quad \frac{N_1}{N_4} = \frac{T_2 \times T_4}{T_1 \times T_3} \qquad \text{...(3)}$$

If R and T = Pitch circle radius and number of teeth of the gear respectively,

$$R_A + R_B = R_C + R_D \text{ and } t_A + t_B = t_C + t_D$$

From the equations, we can determine the number of teeth on each gear for the given center distance, speed ratio and module only when the number of teeth on one gear is considered arbitrarily. The reverted gear trains are used in automotive transmissions, industrial speed reducers and in clocks.

## Problems

An epicyclic gear train is shown in the following figure.

Let us calculate how many revolution does the arm makes when (1) A makes one revolution in clockwise and D makes 1/2 a revolution in the opposite sense and (2) A makes one revolution in clockwise and D remains stationary. The number of teeth in gears A and D are 40 and 90 respectively.

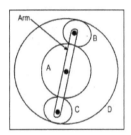

Solution:

Given:

$$T_a = 40$$

$$T_D = 90$$

$$d_A + d_B + d_C = d_D$$

$$d_A + 2d_B = d_D [d_B = d_C]$$

$$T_A + 2T_B = T_O$$

$$2T_B = 90 - 40$$

$$T_B = 25$$

| Conditions of Motion | Revolutions of Elements | | | |
|---|---|---|---|---|
| | Arm | Gear A | Gear B - C | Gear D |
| Arm fixed Gear A rotates through +1 rev. (CCW) | 0 | 1 | $-\dfrac{T_A}{T_B}$ | $-\dfrac{T_A}{T_D}$ |

| | | | | |
|---|---|---|---|---|
| Arm fixed Gear A rotates | 0 | x | $\left\|-x\dfrac{T_A}{T_B}\right\|$ | $\left\|-x\dfrac{T_A}{T_D}\right\|$ |
| Through +x rev. (CCW) Add +y rev. to all elements | y | y | y | y |
| Total Motion | y | x + y | $\left\|y-x\dfrac{T_A}{T_B}\right\|$ | $\left\|y-x\dfrac{\dot{T}_A}{T_D}\right\|.$ |

1. When A makes 1 rev. (CW and D makes 1/2 rev. (CCW)).

$$x + y = 1$$

$$y - x\frac{T_A}{T_D} = 1/2$$

$$y - x\frac{40}{90} = 1/2 \qquad -x + 2.25y = 1.125 \;(\because x + y = 1)$$

$$-40x + 90y = 45 \qquad\qquad 3.25y = 2.125$$

$$-x + 2.25y = 1.125 \qquad\qquad y = \frac{2.125}{3.25}\,(CCW)$$

Speed of arm = y = 0.65 (CCW).

2. When a makes 1 rev. (CW) and D is stationary.

$$x + y = -1$$

$$y - x\frac{T_A}{T_D} = 0$$

$$y - x\frac{40}{90} = 0$$

$$-x + 2.25y = 0 \;\;(\because x + y = -1)$$

$$-y + 2.25y = 0$$

$$3.25y = -1$$

$$y = \frac{-1}{3.25}\,(CW)$$

Speed of arm = y = - 0.308 (CW).

2. An epicyclic gear train is shown in figure. The Input is given to the gear A which has 24 teeth. Gear wheels B and C constitute a compound planet having 30 and 18 teeth

respectively. If all the gears are of the same pitch, let us find the speed ratio of the gear train assuming the gear wheel E is fixed.

Solution:

Given:

$T_a = 24$ teeth; $T_B = 30$ teeth; $T_C = 18$ teeth.

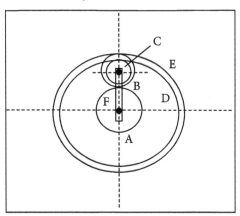

| Sr. No. | Conditions of Motion | Revolutions of Elements | | | | |
|---|---|---|---|---|---|---|
| | | Arm F | Gear A | Compound Gear B-C | Gear D | Gear E |
| 1. | Arm F fixed; Gear A rotates through +1 review.i.e.,(+1 review anticlockwise). | 0 | +1 | $-\dfrac{T_A}{T_B}$ | $-\dfrac{T_A}{T_B} \times \dfrac{T_C}{T_D}$ | $-\dfrac{T_A}{T_B} \times \dfrac{T_B}{T_E} = -\dfrac{T_A}{T_E}$ |
| 2. | Arm F fixed; sun gear A rotates through +x revolutions. | 0 | +x | $-$ | $-x\dfrac{T_A}{T_B} \times \dfrac{T_C}{T_D}$ | $-x\dfrac{T_A}{T_E}$ |
| 3. | Add +Y revolutions to all elements. | +Y | +Y | +Y | +Y | +Y |
| 4. | Total Motion | +Y | X+Y | $y-x\dfrac{T_A}{T_B}$ | $y-x\dfrac{T_A}{T_B} \times \dfrac{T_C}{T_D}$ | $y-x\dfrac{T_A}{T_E}$ |

Ratio of the Reduction Gear:

$$\text{Ratio of the reduction} = \frac{\text{Speed of the input Gear}}{\text{Speed of the Output Gear}} = \frac{N_A}{N_D}$$

From the geometry of the figure, we can write,

$$\frac{d_E}{2} = \frac{d_A}{2} + d_B$$

and $\dfrac{d_D}{2} = \dfrac{d_A}{2} + \dfrac{d_B}{2} + \dfrac{d_C}{2}$ (or) $d_D = d_A + d_B + d_C$

where $d_A$, $d_B$, $d_c$, $d_D$ and $d_E$ are the pitch circle.

The number of teeth are proportional to the pitch circle diameter. So,

$$\frac{T_E}{2} = \frac{T_A}{2} + T_B \Rightarrow \frac{T_E}{2} = \frac{24}{2} + 30$$

$T_e = 84$ and $T_D = T_A + T_B + T_c$

$T_d = 24 + 30 + 18 = 72$

The given conditions are:

Gear wheel E is fixed. So,

$$Y - x \text{—} = 0$$

$$Y - x \times \frac{24}{84} = 0$$

Or,     $Y = \dfrac{2x}{7}$

Speed of input gear A,

$$N_A = x + Y$$

Speed of output gear D,

$$N_D = Y - x \times \frac{T_A}{T_B} \times \frac{T_C}{T_D}$$

$$N_D = Y - x \times \frac{24}{30} \times \frac{18}{72}$$

$$N_D = Y - 0.2 \, x$$

Ratio of reduction gear,

$$\frac{N_S}{N_D} = \frac{x+Y}{Y-0.2x} = \frac{x+\dfrac{2x}{7}}{\dfrac{2x}{7}-0.2x}$$

$$\frac{N_S}{N_D} = \frac{x\left(1+\dfrac{2}{7}\right)}{x\left(2/7-0.2\right)} = 15$$

**Result:**

The speed ratio $\dfrac{N_S}{N_D} = 15$

3. The pitch circle diameter of the annular gear in the epicyclic gear train in figure is 425 mm and the module is 5 mm. When the annular gear 3 is stationary, the spindle A makes one revolution in the same sense as the sun gear 1 for every 6 revolutions of the driving spindle carrying the sun gear.

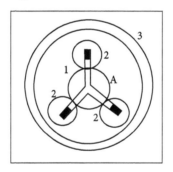

All the planet gears are of the same size. Let us determine the number of teeth on all gears.

Solution:

| Sr. No. | Conditions of Motion | Spider A | Revolutions of Elements | | |
|---|---|---|---|---|---|
| | | | Sun Wheel 1 | Planet Wheel 2 | Annular Gear 3 |
| 1 | Spider A fixed; Sun Wheel rotates through +1 revolve. | 0 | +1 | $\dfrac{T_1}{T_2}$ | $\dfrac{-T_1}{T_2}\times\dfrac{T_2}{T_3}$ $=\dfrac{-T_1}{T_3}$ |

| 2 | Spider A fixed; Sun Wheel 1 rotates through +x revolution. | 0 | +x | $-x\dfrac{T_1}{T_2}$ | $-x\dfrac{T_1}{T_3}$ |
|---|---|---|---|---|---|
| 3 | Add +y revolution to all element. | +y | +y | +y | +y |
| 4 | Total Motion. | +y | x + y | $y-x\dfrac{T_1}{T_2}$ | $y-x\dfrac{T_1}{T_3}$ |

$$y = +1; x + y = +6$$

$$x = 6 - 1 = 5$$

Given gear A is Stationary:

$$y - x\frac{T_1}{T_3} = 0$$

$$1 - 5\frac{T_1}{T_3} = 0$$

$$T_3 = \frac{425}{5} = 85 = T_3$$

$$1 - 5\left(\frac{T_1}{85}\right) = 0$$

$$-T_1 = \frac{-1}{0.0978}$$

$$T_1 = 17$$

$$\frac{d_3}{2} = d_2 + \frac{d_1}{Q}$$

$$d_3 = 2d_2 + d_1$$

$$T_3 = 2T_2 + T_1$$

$$85 = 2T_2 + 17$$

$$d_3 = 2d_2 + d_1$$

$$T_2 = 34$$

### 3.3.3 Epicyclic Train and Their Applications

The gear trains arranged in such a manner that one or more of their members moves upon and around another member are known as epicyclic gear train (epi-means upon and cyclic means around). The epicyclic gear train may be simpler or compound train.

## Epicyclic Gear Train

Epicyclic means one gear revolving upon and around another. The design involves planet and sun gears as one orbits the other like a planet around the sun. Here is a picture of a typical gear box. This design can produce large gear ratios in a small space and are used on a wide range of applications from marine gearboxes to electric screwdrivers.

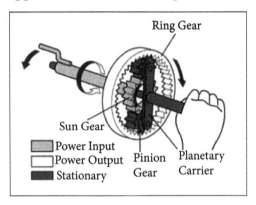

Uses:

- Back Gear of Lathe.

- Differential Gears of Automobile.

- Hoists, Wrist Watches, etc.

## Problem

In an epicyclic gear train shown in figure, the arm A is fixed to the shaft S. The wheel B having 100 teeth rotates freely on the shaft S. The wheel F having 150 teeth driven separately. If the arm rotates at 200 rpm and wheel F at 100 rpm in the same direction; let us determine (a) number of teeth on the gear C and (b) speed of wheel B.

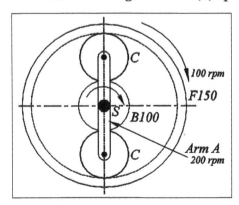

Solution:

Given:

$T_B=100$; $T_F=150$; $N_A=200$rpm; $N_F=100$rpm.

The number of teeth on the gears is proportional to the pitch circles,

$$r_F = r_B+2r_c$$

$$T_F= T_B + 2T_C$$

$$150= 100+2 \times T_c$$

$$T_c= 25 \rightarrow \text{Number of teeth on gear C}$$

The gear B and gear F rotates in the opposite directions,

$$\therefore \text{Train value} = -\frac{T_B}{T_F}$$

$$\text{also,} \quad TV = \frac{N_L - N_{Arm}}{N_F - N_{Arm}} = \frac{N_F - N_A}{N_B - N_A}$$

$$\therefore \quad -\frac{T_B}{T_F} = \frac{N_F - N_A}{N_B - N_A}$$

$$-\frac{100}{150} = \frac{100-200}{N_B - 200} \Rightarrow N_E = 350$$

The Gear B rotates at 350 rpm in the same direction of gears F and Arm A.

## 3.4    Belt, Rope and Chain Drives

### Belt Drive

A belt is a looped strip of flexible material, used to mechanically link two or more rotating shafts. They may be used as a source of motion, to efficiently transmit power or to track relative movement. Belts are looped over pulleys.

In a two pulley system, the belt can either drive the pulleys in the same direction or the belt may be crossed, so that the direction of the shafts is opposite.

### Selection of a Belt Drive

Following are the various important factors upon which the selection of a belt drive depends:

- Speed of the driving and driven shafts.

- Speed reduction ratio.

- Power to be transmitted.

- Centre distance between the shafts.

- Positive drive requirements.

- Shafts layout.

- Space available.

- Service conditions.

The belt drives are usually classified into the following three groups:

1. Light drives: These are used to transmit small powers at belt speeds upto about 10 m/s, as in agricultural machines and small machine tools.

2. Medium drives: These are used to transmit medium power at belt speeds over 10 m/s but up to 22 m/s, as in machine tools.

3. Heavy drives: These are used to transmit large powers at belt speeds above 22 m/s, as in compressors and generators.

## Type of Belt Drive

## Flat Belt

The flat belt, as shown in figure (a), is mostly used in the factories and workshops, where a moderate amount of power is to be transmitted, from one pulley to another when the two pulleys are not more than 8 metres apart.

(a)

## Open Belt Drive

The open belt drive, as shown in figure, is used with shafts arranged parallel and rotating in the same direction. In this case, the driver A pulls the belt from one side (i.e. lower side RQ) and delivers it to the other side (i.e. upper side LM).

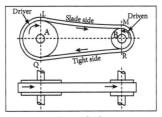

Open belt

Thus the tension in the lower side belt will be more than that in the upper side belt. The lower side belt (because of more tension) is known as tight side whereas the upper side belt (because of less tension) is known as slack side, as shown in figure.

## Crossed or Twist Belt Drive

The crossed or twist belt drive, as shown in figure is used with shafts arranged parallel and rotating in the opposite directions. In this case, the driver pulls the belt from one side (i.e. RQ) and delivers it to the other side (i.e. LM). Thus, the tension in the belt RQ will be more than that in the belt LM. The belt RQ (because of more tension) is known as tight side, whereas the belt LM(because of less tension) is known as slack side, as shown in figure.

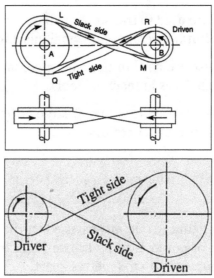

Crossed or twist belt drive.

## V-Belt

The V-belt, as shown in Figure (b), is mostly used in the factories and work-shops, where a moderate amount of power is to be transmitted, from one pulley to another, when the two pulleys are very near to each other.

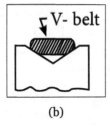

(b)

## Circular Belt or Rope

The circular belt or rope, as shown in Figure (c), is mostly used in the factories and workshops, where a great amount of power is to be transmitted, from one pulley to another, when the two pulleys are more than 8 meters apart.

(c)

Some of the applications of flat belts are given below:

- Drives to beaters on conventional blow rooms.

- Crossed flat-belt transmits drives from cylinder to flat on old cards.

- Drive to drafting rollers and other rolling elements on a single delivery drawing machine.

- Drives to opening rollers, friction drums and take-off rollers on friction spinning machine.

- Drive to rotor on rotor-spinning machine.

- Main drive on draw-texturing machine.

- Drive to creel-rollers of a high speed drawing machine.

## Rope Drive

The rope drives are widely used where a large amount of power is to be transmitted, from one pulley to another, over a considerable distance. It may be noted that the use of flat belts is limited for the transmission of moderate power from one pulley to another when the two pulleys are not more than 8 metres.

If large amounts of power are to be transmitted by the flat belt, then it would result in excessive belt cross-section. One of the main advantage of rope drives is that a number of separate drives may be taken from the one driving pulley. For example, in many spinning mills, the line shaft on each floor is driven by ropes passing directly from the main engine pulley on the ground floor.

The rope drives use the following two types of ropes:

- Fiber ropes.

- Wire ropes.

The fiber ropes operate successfully when the pulleys are about 60 meters apart, while the wire ropes are used when the pulleys are up to 150 meters apart.

## Advantages of Fiber Rope Drives

- The shafts may be out of strict alignment.

- The power may be taken off in any direction and in fractional parts of the whole amount.

- They give high mechanical efficiency.

- They give smooth, steady and quiet service.

- They are little affected by outdoor conditions.

## Belt Material

### Leather

Oak tanned or chrome tanned.

### Rubber

Canvas or cotton duck impregnated with rubber. For greater tensile strength, the rubber belts are reinforced with steel cords or nylon cords.

### Plastics

Thin plastic sheets with rubber layers.

### Fabric

Canvas or woven cotton ducks the belt thickness can be built up with a number of layers. The number of layers is known as ply.

The belt material is chosen depending on the use and application. Leather oak tanned belts and rubber belts are the most commonly used but the plastic belts have a very good strength almost twice the strength of leather belt. Fabric belts are used for temporary or short period operations.

When power is to be transmitted over long distances then belts could not be used due to the heavy losses in power. In such cases ropes will be used. Ropes are used in elevators, mine hoists, cranes, oil well drilling, aerial conveyors, tramways, haulage devices, lifts and suspension bridges etc. two types of ropes are generally used. They are fiber ropes and metallic ropes. Fiber ropes are made of Manila, hemp, cotton, jute, nylon, coir etc. and are usually used for transmitting power. Metallic ropes are made of steel, aluminium. alloys, copper, bronze or stainless steel and are mainly used in elevator, mine hoists, cranes, oil well drilling, aerial conveyors, haulage devices and suspension bridges.

## Hoisting tackle (Block and Tackle Mechanism)

It consists of two pulley blocks one above the other. Each block has a series of sheaves mounted side by side on the same axle. The ropes utilized in hoisting tackle are:

- Cotton ropes

- Hemp ropes

- Manila ropes

The pulleys are manufactured in two designs i.e. fixed pulley and movable pulley.

## Steel Wire Ropes

A wire rope is made up of strands and a strand is made up of one or more layers of wires as shown in figure. The number of strands in a rope denotes the number of groups of wires which are laid over the central core. For example a 6× 19 construction means that the rope has 6 strands and every strand is composed of 19(12/6/1) wires. The central part of the wire rope is known as the core and may be of fiber, wire, plastic, paper or asbestos. The fiber core is very flexible and very suitable for all conditions.

The points to be considered while selecting a wire rope are:

- Strength

- Abrasion resistance

- Flexibility

- Resistance of crushing

- Fatigue strength

- Corrosion resistance

Ropes having wire core are stronger than those having fiber core. Flexibility in rope is more desirable once the number of bends in the rope is too many.

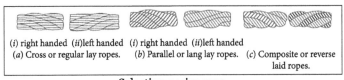

(i) right handed (ii)left handed  (i) right handed (ii)left handed
(a) Cross or regular lay ropes.   (b) Parallel or lang lay ropes.   (c) Composite or reverse laid ropes.

Selecting a wire rope.

## Design Procedure for Wire Rope

Let,

d=Diameter of rope,

D= Diameter of sheave,

H= Depth of mine or height of building,

W= Total load,

$W_R$= Weight of rope,

$D_w$= Diameter of wire,

A= Area of c/s of rope,

$P_b$= Bending load in the rope,

$F_a$= Allowable pull in the rope,

$F_u$= Ultimate of breaking load of rope,

N= Factor of safety,

$W_s$= Starting load,

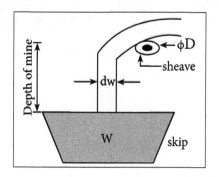

## Total Load

W= Load to be lifted + Weight of skip

## Total Weight of the Load

$W_R$ = Weight per meter length × length of rope

   =Weight per meter length × H

## Inertia Load Due to Acceleration

$$W = Ma = \left( \frac{W \quad W}{} \right).a$$

## Bending Load

$$P_b = \frac{k.Ad_w}{D}$$

Where,

$k = 82728.5\text{Mpa}$ = Modulus of elasticity of rope in $N/mm^2$,

$A$ = Area of c/s of rope in $mm^2$,

$D_w$= Diameter of wire in mm,

$D$ = Diameter of sheave in mm,

## Starting Load

$W_s = 2[W + W_R]$

## Maximum Load

The maximum load on the rope can be determined in the following ways:

- Load on the rope during uniform velocity = $W + W_R + P_b$.
- Load on the rope during acceleration = $W + W_R + W_1 + P_b$.
- Load on the rope during starting = $W_s + P_b$.

$F_{max}$ is the maximum among the above three values.

Neglecting impact $Fr = W + W_R + W_1 + P_b$.

## Diameter of Robe

Allowable pull $F_a \geq F_{max}$

Since $F_a = \dfrac{F_u}{n}$

$\dfrac{F_u}{n} \geq F_{max}$

Neglecting impact,

$$\dfrac{F_u}{n} \geq W + W_R + W_I + P_b$$

$F_u = 500.8$ MN Where d in meters = $500.8 \, d^2 N$ where d in mm.

Weight per unit length of rope = $36.3 \, d^2 kn/m$,

= $36.3d^2 \times 10^{-3}$ N/m where d in mm.

Wire diameter $d_w = 0.063d$ mm.

Area of c/s of rope $A = 0.38d^2 mm^2$, d in mm.

Average sheave diameter D= 45d in mm.

To find the acceleration any one of the following equation may be used,

$$v = u + at \qquad \qquad \qquad \text{...(1)}$$

$$s = ut + 1/2\ at^2 \qquad \qquad \text{...(2)}$$

$$v^2 = u^2 + 2as \qquad \qquad \qquad \text{...(3)}$$

## Problem

1. Let us select a wire rope to lift a load of 10kN through a height of 600m from a mine. The weight of bucket is 2.5kN. The load should attain a maximum speed of 50m/min in 2 seconds.

Solution:

Given:

From table select the most commonly used type of rope i.e. 6×19.

From table for 6×19 rope $F_u$= 500.8 $d^2$ N where d in mm.

Weight per meter length = $36.3 \times 10^{-3}\ d^2$ N/m where d in mm.

Wire diameter $d_w$ = 0.063d, mm.

Area of c/s A = 0.38 $d^2$, mm².

Sheave diameter D= 45 d, mm.

From table for 600 m depth,

F.O.S= n = 7

## Total Load

W= Load to be lifted + weight of skip = 10000+ 2500= 12500N

## Total Weight of Rope

$W_R$ = weight per meter length × length of rope

= $36.3 \times 10^{-3} d^2 \times 600$ N/M where d in mm= $21.78d^2$

## Inertia Load Due to Acceleration

$$W = Ma = \left( \frac{W \quad W}{\quad} \right).a$$

$v = u + at$, since $u = 0$

$V = at$

$a = v/t = 50/(60 \times 2) = 0.417 \, m/sec^2$

$$\therefore W_1 = \left( \frac{12500 + 21.78d^2}{9.81} \right) \times 0.417$$

$= 531.345 + 09258 \, d^2$

## Bending Load

$$P_b = k.A \frac{d_w}{D}$$

Where $k = 82728.5 \, MPa$

$$\therefore P_b = 82728.5 \times 0.38d^2 \times \frac{0.063d}{45d} = 44.01d^2$$

## Starting Load

$W_s = 2[W + W_R] = 2[12500 = 21.78d^2] = 25000 = 43.56d^2$

## Maximum Load

Assume impact load is neglected,

Max. Load on the rope $F_{max} = W + W_R + W_1 + P_b$

$F_a \geq F_{max}$

i.e., $\dfrac{F_u}{n} \geq F_{max}$

$$\therefore \frac{500.8d^2}{7} \geq 13031.345 + 66.7158d^2$$

$d \geq 51.96mm$

$F_{max} = 12500 + 21.78d^2 + 531.345 + 0.9258d^2 + 44.01d^2$

## Diameter of Rope

From table 21.40 for 6x19 rope std diameter d=54mm

2. A belt drive consists of two V-belts in parallel, on grooved pulleys of the same size. The angle of the groove is 30°. The cross-sectional area of each belt is 750 mm² and μ

=0.12.The density of the belt material is 1.2 Mg/m³ and the maximum safe stress in the material is 7 MPa. Let us calculate the power that can be transmitted between pulleys of 300 mm diameter rotating at 1500r.p.m. And also determine the shaft speed in r.p.m. at which the power transmitted would be a maximum.

Solution:

Given:

$n = 2$

$2\beta = 30°$ or $\beta = 15°$

$a = 750$ mm² $= 750 \times 10^{-6}$ m

$\mu = 0.12$

$\rho = 1.2$ Mg/m³ $= 1200$ kg/m³

$\sigma = 7$ MPa $= 7 \times 10^6$ N/m²

$d = 300$ mm $= 0.3$ m

$N = 1500$ r.p.m

To find:

We know that mass of the belt per metre length,

m= Area × length × density = $750 \times 10^{-6} \times 1 \times 1200 = 0.9$ kg/m

Speed of the belt,

$$v = \frac{\pi dN}{60} = \frac{\pi \times 0.3 \times 1500}{60} = 23.56 \, m/s$$

Centrifugal tension,

$$T_c = m \cdot v^2 = 0.9(23.56)^2 = 500 \, N$$
$$T = \sigma \times a = 7 \times 10^6 \times 750 \times 10^{-6} = 5250 \, N$$

We know that tension in the tight side of the belt,

$$T_1 = T - T_c = 5250 - 500 = 4750 \, N$$

$T_2$ = Tension in the slack side of the belt

Since the pulleys are of the same size, therefore angle of lap ($\theta$) = 180° = $\pi$rad.

We know that,

$$2.3 \log\left(\frac{T_1}{T_2}\right) = \mu \cdot \theta \operatorname{cosec} \beta = 0.12 \times \pi \times \operatorname{cosec} 15° = 0.377 \times 3.8637 = 1.457$$

$$\log\left(\frac{T_1}{T_2}\right) = \frac{1.457}{2.3} = 0.6335 \quad \text{or} \quad \frac{T_1}{T_2} = 4.3 \ldots \text{(Taking antilog of 0.6335)}$$

$$T_2 = T_1/4.3$$

$$= 4750/4.3$$

$$= 1105 \text{ N}$$

## Power Transmitted

We know that power transmitted,

$$P = (T_1 - T_2) v \times n (4750 - 1105) 23.56 \times 2 = 171750 \text{ W}$$

$$= 171.75 \text{kW}$$

## Shaft Speed

Let,

$N_1$ = Shaft speed in r.p.m.

$v_1$ = Belt speed in m/s.

We know that for maximum power, centrifugal tension,

$$T_C = T/3 \text{ or } m(v_1)^2 = T/3$$

$$0.9(v_1)^2 = 5250/3 = 1750$$

$$(v_1)^2 = 1750/0.9 = 1944.4 \text{ or } v_1 = 44.1 \text{m/s}$$

We know that belt speed $(v_1)$,

$$44.1 = \frac{\pi d N_1}{60} = \frac{\pi \times 0.3 \times N_1}{60} = 0.0157 N_1$$

$$N_1 = 44.1/0.0157 = 2809 \text{ r.p.m.}$$

## Chain Drive

Chain is used to transmit motion from one shaft to another shaft with the help of sprockets. Chain drives maintain a positive speed ratio between driving and driven components, so tension on the slack side is considered as zero. They are generally used for the transmission of power in cycles, motor vehicles, agricultural machinery, road rollers etc.

## Merits of Chain Drives

- Chain drives are positive drives and can have high efficiency when operating under ideal conditions.

- It can be used for both relatively long centre distances.

- Less load on shafts and compact in size as compared to belt drive.

## Demerits of Chain Drives

- Relatively high production cost and noisy operation.

- Chain drives require more amounts of servicing and maintenance as compared to belt drives. Velocity ratio in chain drive.

Let,

$n_1$= Speed of driver sprocket in rpm.

$n_2$ = Speed of driven sprocket in rpm.

$z_1$= Number of teeth on drivers sprocket.

$z_2$ = Number of teeth on driven sprocket.

Therefore Velocity ratio $n_1/n_2 = z_1/z_2$.

## Chains for Power Transmission

The different types of chain used for power transmission are:

- Block chain.

- Roller chain.

- Inverted-tooth chain or silent chain.

## Roller Chain

It consists of two rows of outer and inner plates. The outer row of plates is known as pin link or coupling link whereas the inner row of plates is called roller link. A Pin passes through the bush which is secured in the holes of the inner pair of links and is riveted to the outer pair of links as shown in Fig. Each bush is surrounded by a roller. The rollers run freely on the bushes and the bushes turn freely on the pins.

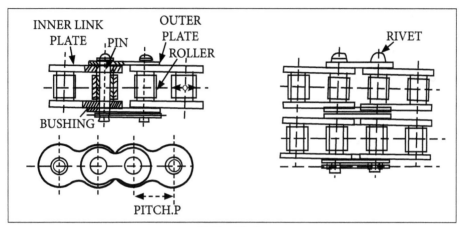

Roller Chain

A roller chain is extremely strong and simple in construction. It gives good service under severe conditions. To avoid longer sprocket diameter, multi-row-roller chains or chains with multiple strand width are used. Theoretically, the power capacity multi strand chain is equal to the capacity of the single chain multiplied by the number of strand, but actually it is reduced by 10 percent.

## Inverted Tooth Chain or Silent Chain

It is as shown figure these chains are not exactly silent but these are much smoother and quieter in action than a roller chain. These chains are made up of flat steel stamping, which make it easy to built up any width desired. The links are so shaped that they engage directly with sprocket teeth. In design, the silent chains are more complex than brush roller types, more expensive and require more careful maintenance.

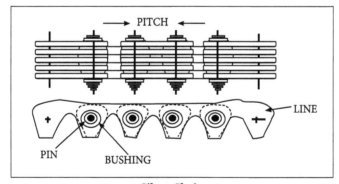

Silent Chain

## Design Procedure for Roller Chain

p = Pitch.

$d_1$ = Diameter of smaller sprocket.

$d_2$ = Diameter of larger sprocket.

$n_1$ = Speed of smaller sprocket.

$n_2$ = Speed of larger sprocket.

$z_1$ = Number of teeth on smaller sprocket.

$z_2$ = Number of teeth on larger sprocket.

L = Length of chain in pitches.

C = Center diameter.

$C_p$ = Center distance in pitches.

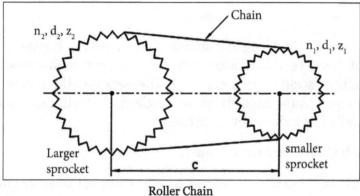

Roller Chain

**Pitch of Chain**

$$p \le 25\left(\frac{900}{n_1}\right)^{\frac{2}{3}}$$

Where p in mm and $n_1$= speed of smaller sprocket.

Select standard nearest value of pitch from Table:

- Chain number.
- Breaking load $F_u$.
- Measuring load W.

**Number of Teeth on the Sprockets**

From Table for the given ratio select the number of teeth on the smaller sprocket ($z_1$). Since,

$$\frac{n_1}{n_2} = \frac{z_2}{z_1}$$

Number of teeth on larger sprocket = $z_2$.

## Pitch Diameters

$$d = \frac{p}{\sin\left(\frac{180}{z}\right)}$$

$$\therefore d_1 = \frac{p}{\sin\left(\frac{180}{z_1}\right)}$$

$$d_2 = \frac{p}{\sin\left(\frac{180}{z_2}\right)}$$

## Velocity

$$v \quad \frac{p z_1 n_1}{60000} \text{ m/sec}$$

## Required Pull

Power required $N = \dfrac{F_\theta \cdot v}{1000 k_1 k_s}$ ----21.115a

$K_t$ = load factor

$\quad$ = 1.1 to 1.5

$K_s$ = service factor

$\quad$ = 1 for 10 hours per day service

$\quad$ = 1.2 for 24 hours operation

## Allowable Pull

Allowable pull $F_a = \dfrac{F_u}{n_o}$

Where $n_o$ = working factor of safety.

## Number of Strands in a Chain

Number of strands $i = \dfrac{F_\theta}{n_a}$

## Check for Actual Factor of Safety

Actual factor of safety $n_a = \left(\dfrac{F_u}{F_\theta + F_{cs} + F_s}\right) i$

Where $F_\theta = \dfrac{1000N}{v}$

$F_{cs} = \dfrac{wv^2}{g}$

$F_s = k_{sg}\, wC$

C = Center distance. If not given for medium center distance $C_p$=30 to 50.

$K_{sg}$ = Coefficient of sag.

i = Number of strands.

If $n_a > n_o$, then safe.

**Allowable Pull**

$$L_p = 2C_p \cos\alpha + \frac{Z_2 + Z_1}{2} + \alpha\left(\frac{Z_2 - Z_1}{180}\right)$$

Where $\alpha = \sin^{-1}\left(\dfrac{d_2 - d_1}{2C}\right)$

**Length of Chain**

$$L = p.L_p$$

**Correct Center Distance**

$$L_p = 2\frac{C}{p}\cos\alpha + \frac{Z_2 + Z_1}{2} + \alpha\left(\frac{Z_2 - Z_1}{180}\right)$$

Correct center distance C.

**Length of Chain**

The open chain drive system connecting the two sprockets is shown in figure. The length of belt for an open belt drive connecting the two pulleys of radii $r_1$ and $r_2$ and a centre distance x is given by,

$$L = \pi(r_1 + r_2) + 2x + \frac{(r_1 + r_2)^2}{x}$$

If this expression is used for finding the length of chain, the result will be slightly greater than the required length. This is due to the fact that the pitch lines A B C D E F G and P Q R S of the sprockets are the parts of a polygon and not that of a circle.

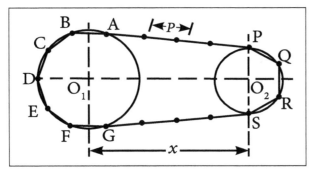

Length of the chain

$T_1$ = Number of teeth on the larger sprocket.

$T_2$ = Number of teeth on the smaller sprocket.

p= Pitch of the chain.

The diameter of the pitch circle,

$$d = p\,cosec\left(\frac{180°}{T}\right) \text{ or } r = \frac{p}{2}cosec\left(\frac{180°}{T}\right)$$

For larger sprocket,

$$r_1 = \frac{p}{2}cosec\left(\frac{180°}{T_1}\right)$$

and for smaller sprocket $r_2 = \frac{p}{2}cosec\left(\frac{180°}{T_2}\right)$

Since the term $\pi(r_1 + r_2)$ is equal to half the sum of the circumferences of the pitch circles, the length of chain is equivalent to,

$$\pi(r_1 + r_2) = \frac{p}{2}(T_1 + T_2)$$

Substituting the values of $r_1$, $r_2$ and $\pi(r_1 + r_2)$ in equation, the length of chain is given by,

$$L = \frac{p}{2}(T_1 + T_2) + 2x + \frac{\left[\frac{p}{2}cosec\left(\frac{180°}{T_1}\right) - \frac{p}{2}cosec\left(\frac{180°}{T_2}\right)\right]^2}{x}$$

If x =m.p, then,

$$L = p\left[\frac{(T_1 + T_2)}{2} + 2m + \frac{\left[cosec\left(\frac{180°}{T_1}\right) - cosec\left(\frac{180°}{T_2}\right)\right]^2}{4m}\right] = p.K$$

K = multiplying factor,

$$= \frac{(T_1+T_2)}{2}+2m+\frac{\left[\operatorname{cosec}\left(\dfrac{180°}{T_1}\right)-\operatorname{cosec}\left(\dfrac{180°}{T_2}\right)\right]^2}{4m}$$

The value of multiplying factor (K) may not be a complete integer. But the length of the chain should be equal to an integer number of times the pitch of the chain. Thus, the value of K should be rounded off to the next higher integral number.

## Advantages of Chain Drive over Belt or Rope Drive

## Advantages

- The chain drive gives less load on the shafts.

- Since the chains are made of metal, therefore they occupy less space in width than a belt or rope drive.

- The chain drive gives high transmission efficiency (up to 98 per cent).

- The chain drives may be used when the distance between the shafts is less.

- As no slip takes place during chain drive, hence perfect velocity ratio is obtained.

- The chain drive has the ability of transmitting motion to several shafts by one chain only.

## Disadvantages

- The chain drive has velocity fluctuations especially when unduly stretched.

- The chain drive needs accurate mounting and careful maintenance.

- The production cost of chains is relatively high.

## Problem

1. Let us Select a roller chain drive to transmit power of 10kw from a shaft rotating at 750 rpm to another shaft to run at 450 rpm. The distance between the shaft centers could be taken as 35 pitches.

Solution:

Given:

$N$= 10 kw; $n_1$ = 750 rpm; $n_2$ = 450 rpm; C = 35 pitches

## 1. Pitch of Chain

$$p \leq 25 \left( \frac{900}{n_1} \right)^{2/3}$$

$$\leq 25 \left( \frac{900}{750} \right)^{2/3}$$

$$\leq 28.23 \, \text{mm}$$

From the nearest standard value of pitch p= 25.4mm.

Select chain number 208 B.

Breaking load $F_u$ = 17.9 KN = 17900 N.

Measuring lad w= 127.5 N.

## 2. Number of Teeth on the Sprockets

$$\frac{n_1}{n_2} = \frac{750}{450} = 1.667$$

From $n_1/n_2$ = 1.667, select number of teeth on the smaller sprocket $z_1$= 27.

Now, $\dfrac{n_1}{n_2} = \dfrac{z_2}{z_1}$

$$\frac{750}{450} = \frac{z_2}{27}$$

Number of teeth on larger sprocket $z_2$ = 45.

## 3. Pitch Diameters

$$d = \frac{p}{\sin \left( \frac{180}{z} \right)}$$

Pitch diameter of smaller sprocket,

$$d_1 = \frac{p}{\sin \left( \frac{180}{z_1} \right)} = \frac{25.4}{\sin \left( \frac{180}{27} \right)}$$

$$= 218.79$$

Pitch diameter of larger sprocket $d_2$,

$$d_2 = \frac{p}{\sin\left(\dfrac{180}{z_2}\right)} = \frac{25.4}{\sin\left(\dfrac{180}{45}\right)}$$

$$= 364.124 \text{mm}$$

## 4. Velocity

$$v = \frac{pz_1 n_1}{60000} = \frac{25.4 \times 27 \times 750}{60000} = 8.57 \text{m/sec}$$

## 5. Required Pull

Power $N = \dfrac{F_\theta \cdot v}{1000 k_1 k_s}$ .....................21.115a (DDHB)

$K_t$ = load factor = 1.1 - 1.5

$K_s$ = service factor

= 1.2 for 24 hours operation (Assume 24 hours operation)

Take kt = 1.3

$$\therefore 10 = \frac{F_\theta \times 8.57}{1000 \times 1.3 \times 1.2}$$

$$F_\theta = 1820.3 \text{N}$$

## 6. Allowable Pull

$F_a = \dfrac{F_u}{n_0}$ where $n_0$ = Working factor of safety.

From $n_1$ = 750 rpm and p=25.4,mm.

Select working factor of safety $n_0$ =11.7 [$n_0$ is not equal to 10.7.]

$$\therefore F_a = \frac{17900}{11.7} = 1529.914$$

## 7. Number of Strands in a Chain

$$i = \frac{F_\theta}{F_a} = \frac{1820.3}{1529.914} = 1.189$$

$\therefore$ Number of strands i = 2

## 8. Check for Actual Factor of Safety

Actual factor of safety:

$$n_a = \left(\frac{F_u}{F_\theta + F_{cs} + F_s}\right)i$$

$$F_0 = \frac{1000N}{v} = \frac{1000 \times 10}{8.57} = 1166.86\,N$$

$$F_{cs} = \frac{wv^2}{g} = \frac{127.5 \times 8.57^2}{9.81} = 954.56\,N$$

$$F_s = k_{sg}wC$$

(for horizontal drive, $k_{sg}$=6),

$$\therefore F_S = 6 \times 127.5 \times \frac{35 \times 25.4}{1000} = 680.085\,N$$

$$\therefore n_a = \left(\frac{17900}{1166.86 + 954.56 + 680.085}\right) \times 2 = 12.778$$

Since $n_a > n_o$, the selection of the chain is safe.

## 9. Length of Chain Pitch

$$L_p = 2C_p\cos\alpha + \frac{z_1 + z_2}{2} + \alpha\left(\frac{z_2 - z_1}{180}\right) \qquad \text{----21.122(DDHB)}$$

$$\alpha = \sin^{-1}\left(\frac{d_2 - d_1}{2C}\right) \qquad \text{----21.122(DDHB)}$$

$$= \sin^{-1}\left(\frac{364.124 - 218.79}{2 \times 35 \times 25.4}\right) = 4.6886$$

$$\therefore L_P = 2 \times 35\cos 4.6886 + \left(\frac{27 + 45}{2}\right) + 4.6886\left(\frac{45 - 27}{180}\right)$$

$$= 106.2346 \text{ pitches}$$

The nearest even number of pitches is 106.

$L_p$ = 106 pitches

## 10. Correct Centre Distance

$$L_p = 2\frac{C}{p}\cos\alpha + \frac{(z_2 + z_1)}{2} + \alpha\frac{(z_2 - z_1)}{180}$$

$$106 = 2 \times \frac{C}{25.4}\cos 4.6886 + \left(\frac{27 + 45}{2}\right) + 4.6886\left(\frac{45 - 27}{180}\right)$$

C=886 mm

2. A chain drive is used for reduction of speed from 240 r.p.m. to 120 r.p.m. The number of teeth on the driving sprocket is 20. Let us determine the number of teeth on the driven sprocket. If the pitch circle diameter of the driven sprocket is 600 mm and centre to centre distance between the two sprockets is 800 mm and let us also determine the pitch and length of the chain.

Solution:

Given:

$N_1$ = 240 r.p.m

$N_2$ = 120 r.p.m

$T_1$ = 20

$d_2$ = 600 mm or

$r_2$ = 300 mm = 0.3 m

x = 800 mm = 0.8 m

To find:

- The number of teeth on the driven sprocket.
- The determine the pitch and length of the chain.

**Number of Teeth on the Driven Sprocket**

Let $T_2$ = Number of teeth on the driven sprocket.

We know that,

$$N_1.T_1 = N_2.T_2 \text{ or } T_2 = \frac{N_1.T_1}{N_2} = \frac{240 \times 20}{120} = 40$$

**Pitch of the Chain**

Let p = Pitch of the chain.

We know that pitch circle radius of the driven sprocket $(r_2)$,

$$0.3 = \frac{p}{2}\text{cosec}\left(\frac{180°}{T_2}\right) = \frac{p}{2}\text{cosec}\left(\frac{180°}{40}\right) = 6.37p$$

$$p = 0.3/6.37 = 0.0471\,m = 47.1\,mm$$

**Length of the chain**

We know that pitch circle radius of the driving sprocket,

$$r_1 = \frac{p}{2}\cosec\left(\frac{180°}{T_1}\right) = \frac{47.1}{2}\cosec\left(\frac{180°}{40}\right) = 150.5\text{mm}$$

$$x = m.p \text{ or } m = x/p = 800/47.1 = 16.985$$

We know that multiplying factor,

$$K = \frac{(T_1+T_2)}{2} + 2m + \frac{\left[\cosec\left(\frac{180°}{T_1}\right)-\cosec\left(\frac{180°}{T_2}\right)\right]^2}{4m}$$

$$= \frac{(20+40)}{2} + 2\times16.985 + \frac{\left[\cosec\left(\frac{180°}{20}\right)-\cosec\left(\frac{180°}{40}\right)\right]^2}{4\times16.958}$$

$$= 30 + 33.97 + \frac{(6.392-12.745)^2}{67.94} = 64.56 \text{ say } 65$$

Length of the chain,

$$L = p.k = 47.1\times65 = 3061\text{mm} = 3.0615\text{m}$$

## 3.4.1 Initial Tension and Effect of Centrifugal Tension on Power Transmission

Let tension on the tight and slack side be '$T_1$' and '$T_2$' respectively. Let $\theta$ be the angle of lap and let $\mu$ be the coefficient of friction between the belt and the pulley. Consider an infinitesimal length of the belt PQ which subtend an angle $\delta\theta$ at the centre of the pulley.

Let 'R' be the reaction between the element and the pulley. Let 'T' be tension on the slack side of the element, i.e. at point P and let $(T + \delta T)$ be the tension on the tight side of the element.

The tensions T and $(T + \delta T)$ shall be acting tangential to the pulley and thereby normal to the radii OP and OQ. The friction force shall be equal to '$\mu R$' and its action will be to prevent slipping of the belt. The friction force will act tangentially to the pulley at the point S.

Consider equilibrium of the element at S and equating it to zero. Resolving all the forces in the tangential direction,

$$\mu R + T\cos\frac{\delta\theta}{2} - (T+\delta T)\cos\frac{\delta\theta}{2} = 0$$

$$\mu R = \delta T\cos\frac{\delta\theta}{2}$$

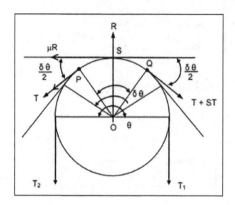

## Centrifugal Tension

The belt continuously runs over the pulleys, there-fore, some centrifugal force is caused, whose effect is to increase the tension on both, tight as well as the slack sides. The tension caused by centrifugal force is called centrifugal tension.

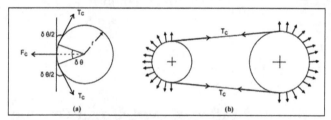

Tension due to centrifugal force.

Let,

'$T_c$' = The centrifugal tension due to centrifugal force.

Let us consider a small element which subtends an angle $\delta\theta$ at the center of the pulley.

'm' = The mass of the belt per unit length of the belt in kg/m'.

Resolving all the forces in the radial direction at S and equating it to zero.

$$R - T\sin\frac{\delta\theta}{2} - (T + \delta T)\sin\frac{\delta\theta}{2} = 0$$

$$R = (2T + \delta T)\sin\frac{\delta\theta}{2}$$

Since $\delta\theta$ is very small, taking limits,

$$\cos\frac{\delta\theta}{2} = 1 \text{ and } \sin\frac{\delta\theta}{2} = \frac{\delta\theta}{2}$$

$$R = (2T + \delta T)\frac{\delta\theta}{2} = T\delta\theta + \delta T\frac{\delta\theta}{2}$$

Neglecting the product of the two infinitesimal quantities:

$\left(\delta T \dfrac{\delta \theta}{2}\right)$ Which is negligible in comparison to other quantities:

$R = T\delta\theta$

Substituting the value of R and $\cos\dfrac{\delta\theta}{2} = 1$

$\mu T \delta\theta = \delta T$

We get, $\dfrac{\delta T}{T} = \mu\delta\theta$

Taking limits on both sides as $\delta\theta \to 0$,

$\dfrac{dT}{T} = \mu d\theta$

Integrating between limits, it becomes,

$\displaystyle\int_{T_2}^{T_1} \dfrac{dT}{T} = \int_{0}^{\theta} \mu\, d\theta$

$In \dfrac{T_1}{T_2} = \mu\theta$

$\dfrac{T_1}{T_2} = e^{\mu\theta}$

The centrifugal force '$F_c$' on the element will be given by,

$F_C = (r\,\delta\theta\, m) \times \dfrac{V^2}{r}$

Where V is speed of the belt in m/sec. And r is the radius of pulley in m.

Resolving the force on the elements normal to the tangent,

$F_C - 2T_C \sin\dfrac{\delta\theta}{2} = 0$

Since $\delta\theta$ is very small,

$\therefore \qquad \sin\dfrac{\delta\theta}{2} = \dfrac{\delta\theta}{2}$

Or, $F_C - 2T_C \dfrac{\delta\theta}{2} = 0$

Or, $F_C = T_C\, \delta\theta$

Substituting for $F_c$,

$$\frac{mV^2}{r} r\, \delta\theta = T_c\, \delta\theta$$

$$\text{Or, } T_c = mV^2$$

Therefore considering the effect of the centrifugal tension, the belt tension on the tight side when power is transmitted is given by,

Tension of tight side $T_t = T_1 + T_c$ and tension on the slack side $T_s = T_2 + T_c$.

The centrifugal tension has no effect on the power transmitted because maximum tension can by only $T_t$ which is,

$$T_t = \sigma_t \times t \times b$$

$$T_1 = \sigma_t \times t \times b\text{-}mV^2$$

## 3.4.2 Maximum Power Transmission Capacity

## Power Transmitted by Belt Drive

Power transmitted by the belt, $P = (T_1 - T_2)v$,

$$\therefore P = \left[ T_1 - \frac{T_1}{e^{\mu\theta}} \right] v = T_1 v \left[ 1 - \frac{T_1}{e^{\mu\theta}} \right]$$

$$= [T - T_c] v \left[ 1 - \frac{1}{e^{\mu\theta}} \right] = (T - mv^2) v \left[ 1 - \frac{1}{e^{\mu\theta}} \right]$$

To obtain the condition for maximum power transmission,

$$\frac{dP}{dv} = 0$$

$$\frac{d}{dv} \left[ (T - mv^2) v \left( 1 - \frac{1}{e^{\mu\theta}} \right) \right] = 0$$

$$\left( 1 - \frac{1}{e^{\mu\theta}} \right) \frac{d}{dv} (Tv - mv^3) = 0$$

$$T - 3mv^2 = 0 \text{ or } v = \sqrt{\frac{T}{3m}}$$

$$T - 3T_c = 0 \text{ or } T_c = \frac{T}{3}$$

To calculate the limiting values T and [T] can be equated. Therefore, for maximum transmission of power, centrifugal tension in the belt should be 1/3rd of the maximum permissible tension in the belt or velocity of the belt should be be $\sqrt{\dfrac{[T]}{3m}}$.

## 3.4.3 Belt Creep

Consider an open belt drive rotating in clockwise direction as shown in the figure. The portion of the belt leaving the driven and entering the driver is known as tight side and portion of belt leaving the driver and entering the driven is known as slack side.

During rotation there is an expansion of belt on tight side and contraction of belt on the slack side. Due to the uneven contraction and expansion of the belt over the pulleys, there will be a relative movement of the belt over the pulleys, which is known as creep in belts.

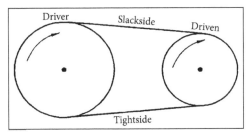

The effect of creep can be reduced by slightly varying the speed of driven pulley or follower.

Velocity ratio is given by,

$$\frac{N_2}{N_1} = \frac{d_1}{d_2} \times \frac{E + \sqrt{\sigma_2}}{E + \sqrt{\sigma_1}}$$

$\sigma_1$ and $\sigma_2$ =Stress in the belt on the tight and slack side respectively.

E= young's modulus for the material of the belt.

## Problem

1. The power is transmitted from a pulley 1 m diameter running at 200r.p.m. to a pulley of 2.25 m diameter by means of a belt. Let us determine the speed lost by the driven pulley as a result of creep, if the stress on the tight and slack side of the belt is 1.4 MPa and 0.5 MPa respectively. The Young's modulus for the material of the belt is 100 MPa.

Solution:

Given:

$d_1 = 1$ m

$N_1 = 200$ r.p.m.

$d_2 = 2.25$ m

$\sigma_1 = 1.4$ MPa $= 1.4 \times 10^6 \text{N/m}^2$

$\sigma_2 = 0.5$ MPa $= 0.5 \times 10^6 \text{N/m}^2$

E= 100 MPa = $100 \times 10^6 \text{N/m}^2$

To find:

The speed lost by the driven pulley as a result of creep,

$N_2$= Speed of the driven pulley.

Neglecting creep, we know that,

$$\frac{N_2}{N_1} = \frac{d_1}{d_2} \text{ or } N_2 = N_1 \times \frac{d_1}{d_2} = 200 \times \frac{1}{2.25} = 88.9 \text{ r.p.m}$$

Considering creep, we know that,

$$\frac{N_2}{N_1} = \frac{d_1}{d_2} \times \frac{E + \sqrt{\sigma_2}}{E + \sqrt{\sigma_1}}$$

$$\text{Or } N_2 = 200 \times \frac{1}{2.25} \times \frac{100 \times 10^6 + \sqrt{0.5 \times 10^6}}{100 \times 10^6 + \sqrt{1.4 \times 10^6}} = 88.7 \text{ r.p.m}$$

Speed lost by driven pulley due to creep,

$$= 88.9 - 88.7 = 0.2 \text{ r.p.m.}$$

### 3.4.4 Slip in Belt

Consider an open belt drive rotating in clockwise direction, the rotation of belt over the pulleys is due to firm frictional grip between the belt and pulleys. When this frictional grip becomes insufficient, there is a possibility of forward motion of driver without carrying belt with it and there is also possibility of belt rotating without carrying the driver pulley with it. This phenomenon is is known as slip.

Therefore slip may be defined as the relative motion between the pulley and the belt in it. The velocity ratio is reduced during slip.

#### Effect of Slip on Velocity Ratio

Let $S_1$=ercentage of slip between driver pulley rim and the belt. $S_2$=Percentage of slip between the belt and the driven pulley rim.

Liner speed of driver $=\pi d_1 n_1$

$$\therefore \text{Linear speed of belt} = \pi d_1 n_1 - \frac{\pi d_1 n_1 s_1}{100} = \pi d_1 \left(1 - \frac{s_1}{100}\right)$$

$$\text{Hence, speed of driven} = \pi d_1 n_1 \left(1 - \frac{S_1}{100}\right)\left(1 - \frac{S_2}{100}\right)$$

i.e., $\pi d_2 n_2 = \pi d_1 n_1 \left(1 - \dfrac{S_1}{100}\right)\left(1 - \dfrac{S_2}{100}\right)$

$\therefore$ Velocity ratio $\dfrac{n_1}{n_2} = \dfrac{d_2}{d_1\left(1 - \dfrac{S_1}{100}\right)\left(1 - \dfrac{S_2}{100}\right)}$

$= \dfrac{d_2}{d_1\left(\dfrac{1 - S_1 + S_2}{100}\right)} = \dfrac{d_2}{d_1\left(1 - \dfrac{S}{100}\right)}$

Neglect $\dfrac{S_1.S_2}{10000}$ since very small

Where s = Total percentage slip = $S_1 + S_2$

Considering thickness velocity ratio $\dfrac{n_1}{n_2} = \dfrac{d_2 + t}{(d_1 + t)\left(1 - \dfrac{S}{100}\right)}$

**Problem**

An engine, running at 150 r.p.m., drives a line shaft by means of a belt. The engine pulley Is 750 mm diameter and the pulley on the line shaft being 450 mm. A 900 mm diameter pulley on the line shaft drives a 150 MN² diameter pulley keyed to a dynamo shaft. Let us determine the speed of the dynamo shaft, when there is no slip and there is a slip of 2% at each drive.

Solution:

Given:

$N_1$ = 150 r.p.m.

$d_1$ = 750 mm

$d_2$ = 450 mm

$d_3$ = 900 mm

$d_4$ = 150 mm

The arrangement of belt drive is shown in figure,

Let, $N_4$ = Speed of the dynamo shaft.

To find:

1. There is no slip.

2. There is a slip of 2% at each drive.

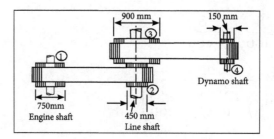

When there is no slip:

We know that $\dfrac{N_4}{N_1} = \dfrac{d_1 \times d_3}{d_2 \times d_4}$ or $\dfrac{N_4}{150} = \dfrac{750 \times 900}{450 \times 150} = 10$

$N_4 = 150 \times 10 = 1500 \, \text{r.p.m}$

## When There is a Slip of 2% at Each Drives

$$\frac{N_4}{N_1} = \frac{d_1 \times d_3}{d_2 \times d_4}\left(1 - \frac{S_1}{100}\right)\left(1 - \frac{S_2}{100}\right)$$

$$\frac{N_4}{150} = \frac{750 \times 900}{450 \times 150}\left(1 - \frac{2}{100}\right)\left(1 - \frac{2}{100}\right) = 9.6$$

$N_4 = 150 \times 9.6 = 1440 \, \text{rpm}$

# Permissions

# Index

**A**

Absolute Velocity, 24-25

Acceleration, 24-25, 27-31, 46, 49-51, 53-63, 65-67, 70, 84, 89, 91, 131, 156-158

Acceleration Analysis, 25

Angular Displacement, 24, 79

**B**

Blocks, 132-134, 155

Brake Power, 133-136, 138

**C**

Centre of Gravity, 28, 56, 75-79, 81, 84

Circular Pitch, 142

Coefficient, 87, 93-94, 96, 98-99, 101, 109, 113, 115, 119, 123, 129, 166, 173

Coefficient of Friction, 96, 98-99, 101, 109, 113, 115, 119, 123, 129, 173

Combustion, 86-87

Configuration, 35, 38-39, 41, 45, 48-49, 60, 66

Connecting Rod, 11, 14, 26-28, 38, 41-42, 48, 51, 53-60, 66, 68-72, 74, 77, 82-84

Constant, 21, 27, 46, 60, 72-74, 102, 133-134

Constant Pressure, 73

Corrosion Resistance, 155

Crank Chain, 17-18, 20, 55

Crank Shaft, 27, 71-72, 86

**D**

Degree of Freedom, 8-10, 13

Dependent Variable, 25

Diameter, 71-72, 96-99, 113-114, 116, 127, 129-132, 134-135, 138, 146-147, 155-160, 163-164, 167, 169-170, 172, 177

**E**

Equilibrium, 106, 120-121, 173

Exhaust, 73, 86-87

Exhaust Stroke, 86-87

Expansion Stroke, 72-73, 86-87

**F**

Fatigue, 155

Flexible Link, 4

Fluid Link, 4

Flywheel, 81, 86-87, 89-90, 92, 94-95, 109-111, 127-129, 134

Flywheels, 92

Four Bar Mechanism, 15, 25, 40, 107

Friction, 95-96, 98-103, 105-113, 115-116, 118-120, 122-125, 127, 129, 132, 138, 153, 173

**G**

Gas Pressure, 16

Gear Box, 149

**K**

Kinetic Energy, 120, 128

Kutzbach Criterion, 8-9

**L**

Linear Velocity, 24, 27-28, 40, 43-44, 47, 60, 63

Loops, 87

**M**

Maximum Pressure, 113, 115, 117

Mean Diameter, 96, 99, 113-114, 116

Means, 4, 42, 65, 77, 86, 101, 110, 119, 132, 138, 149, 155, 177

Moment of Inertia, 75, 80-82, 84, 92

Motion, 1-12, 14-17, 19, 21-22, 25, 29-32, 38, 40, 44, 46, 51, 60-61, 75, 79, 96-97, 105-108, 119, 133, 140-141, 143-144, 147-148, 150, 161, 168, 178

**O**

Odd Number, 140

Oscillations, 79, 83

Outer Radius, 115

**Q**

Quick Return, 11-12, 22, 46, 61

**R**

Radius, 21, 26, 36, 42, 51, 56, 58, 67, 76-77, 81, 84, 87-90, 92, 94, 99, 101-104, 106, 112, 115, 117, 119, 123-124, 127, 133, 135, 142-143, 172, 175

Reciprocating Engine, 18, 51, 58-59

Relative Motion, 1-2, 4-5, 17, 38, 178

Resistance, 67, 71, 96-97, 101, 119, 132-133, 155

Resonant Frequency, 78

Rigid Body, 10, 32-33, 75-76, 84

Rigid Link, 4, 17, 33

Rotation, 6, 25, 31, 33, 40, 46, 56, 79, 105-106, 119, 121-122, 138, 140, 177-178

**S**

Shear Stress, 129

Simple Gear Train, 139-140

Sliding Pair, 2

**T**

Tensile Strength, 154

Theorem, 38-39, 47, 49, 52, 81

Top Dead Centre, 54

Translation, 6, 40

Transmit Power, 150, 168

Transmitting Motion, 4, 168

Turning Pairs, 21, 106

**V**

Velocity, 21, 24-30, 32-49, 51-55, 60-66, 100, 102, 129, 139, 141, 157, 162, 165, 168, 170, 176-178

Velocity Vector, 38, 41, 45, 60

**Z**

Zero Velocity, 24

Printed in the USA
CPSIA information can be obtained
at www.ICGtesting.com
JSHW051621061123
51533JS00005B/51

9 781647 254278